The Atemporal Particle Theory

New supporting evidence

Includes review and commentary of:
Jeff Hawkins' brilliant book
On Intelligence
and Dr. Francis Crick's
The Astonishing Hypothesis

by John Beiswenger

Foreword

John has uncovered a theory that explains the paradox of body, spirit and soul. His research into unexplained phenomena, and his integration of the current thinking of many famous intellectuals with the Atemporal Particle Theory are very convincing. Read this book with an open mind. Take time to understand the theory and report any evidence you see in your own life to corroborate it. As John outlines the 3 states of existence--Temporal, Sequential and Concurrent--you may see the obvious continuity of his theory in explaining reality for the past, present and future from a Christian perspective.

James A Wilson, MD Lancaster PA

Contents

Section	Page
Foreword	2
Preface	4
1 Unanswered Questions	8
2 Unexplained Phenomena	12
3 Hypothetical Process of Research	19
4 Dr. Crick's *Astonishing Hypothesis*	22
5 Jeff Hawkins' *On Intelligence*	26
6 The Foundational Concept	30
7 Creation	35
8 **The Atemporal Particle Theory**	40
9 Evidence of the Incorporeal Body	44
10 Evidence of the Atemporal Particle	59
11 Memory and Consciousness	63
12 Action Potentials	82
13 Reaction to Unexpected Events	89
14 Reproduction	94
15 Life and Healing	99
16 Aging	112
17 Death – Transition	116
18 Conclusions	126
19 Layperson's Addendum	139
20 Bibliography	144
21 Index	147
22 End Notes begin	156
Catholic Catechism compared	174
Cancer and the Atemporal Particle Theory	180
Afterword	185
Acknowledgements	187

Preface

As a native of Milwaukee, Wisconsin, I studied at Marquette University School of Engineering and at the University of Wisconsin, Whitewater, where I majored in Physics. However, I am a scientist by definition and experience, not by education.

I started my first company to design and manufacture electronic products when I was 23, and I sold the company, and the rights to my first patent, a year later to an automotive accessory manufacturer, where I was hired as an electronics design engineer. I then joined a professional engineering firm and worked as a product research consultant to manufacturers when I was 25.

I now have over 50 years of experience in product research, design engineering, product development, manufacturing, product management, general management, marketing and sales in high volume consumer and commercial hard goods.[1] Half of my career was spent working for manufacturers and half as

---

[1] My resume is available through www.beiswenger.com

a consultant to manufacturers. I was responsible for nearly 65 products going into production, personally monitoring pilot production in the U.S., Ireland, France, Japan, Taiwan, Hong Kong, and P.R China. I am named on over 30 issued U.S. and foreign utility patents, the most recent including:

- Medical device for predictive health monitoring,
- Early warning system for bioterrorism events,
- Color LCD touch-display technology,
- Fingerprint scanning technology and
- Surgical instrument sterilization.

In 2007 I co-founded a medical-device company with three others. The company had patented technology, which a registered nurse, an MD and I developed, capable of detecting respiratory infections, such as influenza, bronchitis and pneumonia, in monitored individuals before they experience symptoms or signs of the illness, such as a fever, permitting early medical intervention.

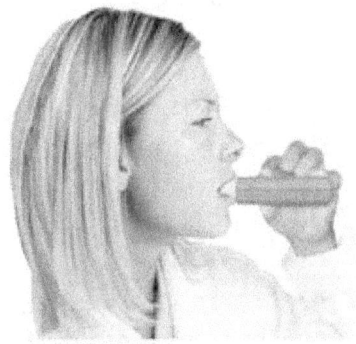

The technology is capable of monitoring individuals or entire populations for the detection and control of pandemics and bioterrorism attacks.

The product and operating algorithm were licensed to a medical device company in 2019. I was appointed Chief Technology Officer of the company.

Thirty-seven years ago, I was intrigued by two seemingly disconnected questions:

> Question 1. How can enough information be stored in a zygote [2] to become a living, breathing human being?

I wrote to scientist, Carl Sagan, commenting on an article of his, and said that he surely knew that no scientist truly understood how so much information could possibly be stored in the human cell. He answered:

> "Natural selection is precisely a process which can extract an enormous amount of information from chaos - including the amount of information in the gene. This works, of course, only if life has been around for a few billion years."

It seems scientists today must mutter such absurdities while they hide in their Darwinian caves, hoping the rocks will fall on their heads rather than face the existence of a Creator.

> Question 2. How can we recall a memory of an event or place, in detail, stored in our "mind" fifty years ago or more and do it in an instant since the brain appears to be too slow to accomplish this?

My wife, Kim, and I were on vacation at the time I first considered these two questions. We were staying in our motorhome along a Lake Michigan beach in Wisconsin.

---

[2] a diploid cell resulting from the fusion of two haploid gametes; a fertilized ovum.

I had just confirmed a rather exciting discovery of mine regarding the luteal phase of women who typically lose their pregnancies at the start of menses. I phoned in my findings from a pay telephone near the beach (1983) to a research physician at the Medical College of Wisconsin and found that I had correctly identified their control subject as having the discovered characteristic, picking her basal metabolic temperature [3] (BMT) graph out of those of the 25 participants in the study.

When you complete a piece of work it has the effect of clearing your mind, freeing it for a new subject of investigation. I began my research regarding the two questions on that day.

---

[3] Basal Metabolic Temperature (BMT), taken upon waking, tracks basal metabolism quite accurately.

Section 1
Unanswered Questions

Scientists will not be able to fully understand vision, consciousness, autism, savant abilities, nor cellular control, including DNA replication, mitosis, meiosis, and certainly not cancer, until scientists look beyond the corporeal aspect of life.

Francis H. C. Crick, OM, FRS, was a British molecular biologist, biophysicist, and neuroscientist, most noted for being a Nobel Prize winner and co-discoverer of the structure of DNA. In his book,[4] Dr. Crick wrote, "Most religions hold that some kind of spirit exists that persists after one's bodily death and, to some degree, embodies the essence of that human being." Crick developed a hypothesis, which can be summed up just as he proposed: "As Lewis Carroll's Alice might have phrased it: 'You're nothing but a pack of neurons,'" he said.

---

[4] Francis H. C. Crick, *The Astonishing Hypothesis*, 1994

Section 1 - Unanswered Questions

Dr. Crick went on to conclude, "If the scientific facts (gathered by the scientific community as prompted by my hypothesis) are sufficiently striking and well established, and if they support (my hypothesis), then it will be possible to argue that the idea that man has a disembodied soul is as unnecessary as the age-old idea that there was a Life Force. This," he arrogantly adds, "is in head-on contradiction to the religious beliefs of billions of human beings alive today."

And billions there are. Muslims, Christians and Hindus believe in a soul, totaling approximately 4.5 billion, and other followers of religions believe there is something beyond the corporeal body which is generally referred to as the soul, yet religious leaders have not reached, nor are they reaching for, a consensus of thought clearly defining the soul. Some Christians, believe humans are tripartite beings, i.e., they have a body, soul and spirit, but there is complete confusion on the definition, purpose and difference between the soul and spirit. Other religions, including Catholicism, claim humans are only bipartite, i.e., body and soul beings. [Take notice of how the Atemporal Particle Theory conflicts with the Catholic Catechism,[5] but yet is in agreement with the Bible.] Most religious faiths incorporate unsubstantiated and often contradictory explanations of the spirit and/or soul components of the human.

---

[5] See "The Catholic Catechism compared to the Atemporal Particle Theory" and the End of the End Notes section.

Section 1 - Unanswered Questions

Greek language definitions take the confusion one step further. "Soul (psuchē) is from the word breath i.e., by implication spirit, thus distinguished from the rational and immortal soul." And "Spirit (pněuma) is a current of air, i.e., breath or figuratively a spirit, i.e, the rational soul." Greek definitions include both words in the definition of each word.

Jeffrey Hawkins was elected as a member of the National Academy of Engineering "for the creation of the hand-held computing paradigm and the creation of the first commercially successful example of a hand-held computing device." He has since turned to work in the field of neuroscience and founded the Redwood Center for Theoretical Neuroscience. Hawkins published a brilliant book, *On Intelligence* [6] describing his memory-prediction framework-theory of the brain.

Both Francis Crick and Jeff Hawkins are at a disadvantage in their research regarding the brain and memory. They are both professed atheists, and their bias would not normally permit them to consider the possibility that humans have an atemporal [7] component, and as a result, they both left us with major unanswered questions.

Let us now look at some unexplained phenomena that I believe are surprisingly connected and explained by the Atemporal Particle Theory. The two seemingly unconnected questions on which I began my study, i.e.,

---

[6] Jeff Hawkins, *On Intelligence*, 2004
[7] atemporal – Timeless; outside of or apart from time. *Funk & Wagnalls New International Dictionary*

Section 1 - Unanswered Questions

how can enough information be stored in a zygote to become a human being, and how can we recall a memory of an event or place, in detail, stored in our "mind" fifty years ago, are not disconnected at all. They have the same answer, and so do all the following.

Section 2
Unexplainable Phenomena

The abilities of autistic savants

The corporeal brain cannot possibly perform the remarkable abilities of autistic savants which have been documented thousands of times, going back to 1887 when the condition was first described, with its first mention in scientific literature as early as 1789. The corporeal brain is simply too slow.

How can the savant's brain do what it apparently does?

Dr. Francis Crick's "Binding Problem"

The "Binding Problem," which obsessed Dr. Francis Crick, is defined by him in his book, *The Astonishing Hypothesis*, as the problem of how (a set of) neurons temporarily become active as a unit. He says,

"As an object seen is often also heard, smelled or felt, this binding must also occur across sensory modalities. Our experience of perceptual unity suggests that the

Section 2 – Unexplainable Phenomena

brain in some way binds together, in a mutually coherent way, all those neurons actively responding to different aspects of a perceived object."

What binds the neurons together in this manner? Others wonder also. See End Notes. [i]

Consciousness

There is, perhaps, no central human experience more discussed and misunderstood than the experience of consciousness. Here are excerpts from an article [8] on consciousness:

"Explaining the nature of consciousness is one of the most important and perplexing areas of philosophy, but the concept is notoriously ambiguous. For instance, how is the conscious mental state related to the body? Can consciousness be explained in terms of brain activity? What makes a mental state be a conscious mental state? The problem of consciousness is arguably the most central issue in current philosophy of mind and is also importantly related to major traditional topics in metaphysics, such as the possibility of immortality and the belief in free will."

The question remains, what produces the conscious experience?

---

[8] Internet Encyclopedia of Philosophy – Consciousness

Section 2 – Unexplainable Phenomena

Phantom limb phenomena

Approximately 60% to 80% of individuals with an amputation experience sensation in their missing limbs. To many amputees, phantom limbs seem to move as if they were still there.

"That's due to the brain's memory of the lost limb," some scientists say; but a 16-year old, with congenital absence of the left arm below the elbow, received a minor injury to the stump of her upper arm and subsequently developed a full-length phantom arm, hand and fingers; reported as Case 1 in a medical publication.[9] Therefore, memory could not be the cause of the phantom limb, since there was, in her case, no memory of the missing limb.

What is it the amputees are experiencing?

Unexplained component of the cellular control process

In the book, Molecular Biology of the Cell. 4th edition,[10] the authors, Bruce Alberts et al, did an excellent job of explaining the central components of the cell-cycle control system as being . . .

---

[9] *Phantom Limb Experiences in Congenital Limb-Deficient Adults* by Dr. Ronald Melzack, Department of Psychology, McGill University.
[10] Molecular Biology of the Cell, 4th edition, Bruce Alberts, Alexander Johnson, Julian Lewis, Martin Raff, Keith Roberts, and Peter Walter. New York: Garland Science; 2002.

Section 2 – Unexplainable Phenomena

"Based on Cyclically Activated Protein Kinases (the title of their article). At the heart of the cell-cycle control system, is a family of protein kinases known as cyclin-dependent kinases (Cdks). The activity of these kinases rises and falls as the cell progresses through the cycle."

However, the authors do not even suggest what instructs the cyclin-dependent kinases?

Verifiable near-death-experiences

There have been thousands of reported near-death-experiences (NDEs) in which an individual is medically confirmed to be dead, has an out-of-body experience and is subsequently resuscitated. Many of these NDEs have been verified by scientists and some experienced by scientists. There is only one logical explanation for these events which most scientists want to avoid. The individual having a near-death-experience is clearly in a state, unlimited by time, limited only by the sequence of events, able to pass through walls and travel great distances in an instant.

How is that possible?

Transfer of memories from a donor to recipients

Dr. Paul Pearsall, PhD, author of The Heart's Code,[11] has recorded dozens of cases in which the recipient of a transplanted heart also received some of the memories

---

[11] The Heart's Code: Tapping the Wisdom and Power of Our Heart Energy, Paul Pearsall, Gary E. Schwartz (Foreword by), Linda G. Russek, 1999

of the donor. How can this transfer occur? Dr. Pearsall's book includes one case where an eight-year old girl received the heart of a murdered child. She was able to describe the killer, the weapon, the place, the clothes he wore and what the little girl had said to him. Everything was found to be accurate, and the killer was convicted.

Memory stored in the heart muscle?

Intuition and the subconscious mind

"Intuitive knowing has been behind almost every human activity in which the acquisition of new knowledge and understanding play a significant part. Intuition is man's communication link between his inner and outer minds.[12] It bridges this ever-present gap. For many centuries before this present scientific era, intuitive knowledge was recognized and accepted by many thinkers, who speculated upon its source. Where does (intuitive) knowledge come from? Every major religion and philosophy have their own obscure term for it. These names are convenient, but they are not informative, for they do not tell us what or where is the source, except something invisible and vague outside of ordinary waking consciousness. Whatever it might be, it is certainly too large and long lasting to fit into a finite brain, and must therefore be abstract, i.e., non-physical, yet occasionally accessible to an inquiring mind." [13]

Where does intuitive information come from?

---

[12] "inner and outer minds"? What is being suggested?
[13] http://appliedintuition.net/

Section 2 – Unexplainable Phenomena

How TMS [14] can induce savant-like abilities

"Low-frequency rTMS (repetitive transcranial magnetic stimulation) temporarily inhibits neural activity in a localized area of the cerebral cortex, thereby creating 'virtual lesions'" (Hilgetag et al. 1999; Walsh & Cowey 2000; Hoffman & Cavus 2002; Steven & Pascual-Leone 2006).

"The left anterior temporal lobe (LATL) is implicated in the savant syndrome for both autistic savants as well as savants who emerge late in life as a result of frontotemporal lobe dementia" (Miller et al. 1998, 2000; Hou et al. 2000).

Accordingly, it was predicted (Snyder, in Carter 1999) and subsequently shown (Snyder et al. 2003, 2006; Young et al. 2004; Gallate et al. 2009) [vii] that savant-like skills can sometimes be artificially induced in normal healthy individuals by inhibiting part of the brain (the LATL). This is consistent with the notion that autistic savants have some atypical left-brain dysfunction or inhibition together with right brain compensation (Miller et al. 1998; Treffert 2005; Sacks 2007)." [15]

Turning off part of the corporeal brain turns on part of the corporeal brain? Not logical. So, what is happening?

The recall of distant memories in an instant

If the brain is so slow, as both Francis Crick and

---

[14] transcranial magnetic stimulation
[15] http://rstb.royalsocietypublishing.org/content/364/1522/1399

Section 2 – Unexplainable Phenomena

Jeff Hawkins believe, how can we recall distant memories, in detail, instantly? I can recall teaching my youngest daughter to ride a bike over 50 years ago. I can see her clearly in my mind after I released my grip on her bike and she peddled down the sidewalk in her light-blue outfit only to collapse on the lawn on the right in tears. How can I do that?

Blindsight

"Blindsight is the ability of people who are cortically blind due to damage to their primary visual cortex to respond to visual stimuli that they do not physically see." (Wikipedia – "Blindsight") How is this possible?

Consciousness of vegetative-state patients

"Up to 20% of patients, thought to be in the vegetative state, respond to sound and visual stimuli when their brain is monitored though Electroencephalography (EEG), a technique that measures changes in the brain's electrical activity in response to a variety of stimuli, including verbal instructions." [16]

No physical brain activity (no fMRI response), yet there is electrical activity? An explanation is needed.

To answer these unexplained phenomena, I propose that the scientific community consider the **Atemporal Particle Theory**.

---

[16] *Into the Gray Zone* by Dr. Adrian Owen, June 20, 2017

Section 3
Hypothetical Process of Research

While it is true, the scientist must normally consider only observable, empirical and measurable evidence and apply known principles of reasoning to their search for answers to heretofore unanswered questions, it is also a perfectly good scientific method of research to observe a phenomenon (whether it be objective or subjective), form a hypothesis, observe empirical evidence that supports the hypothesis and then to form a tentative conclusion, i.e., a testable theory, to see if the unexplained can in fact be explained, doing so with an open mind not encumbered by biases.

Here is the approach described by an anonymous physicist talking about quantum physics: "Quantum mechanical laws are rules for the universe, and in that sense they're no weirder than gravity or anything that Newton did. And we explore them in the same way: follow clues, come up with models, test them out, find out that practically all of them are wrong, etc."

And, here is an example of how I used a hypothesis, during a product research project, to help solve an

Section 3 – Hypothetical Process of Research

engineering problem: We needed to design a touch-screen control, for a machine, using the new LCD touch-screen technology we created. The control had to be able to display green for go, red for stop and amber for caution. It was too expensive back then to consider using a full color LCD.

Using cross-aligned color polarizers to filter the light, we could get green and red; the ON state of the LCD gave us green and the OFF state gave us red. But, we asked, how can we get a third color, amber, with a two-state LCD – when segments of the LCD can just be turned on and off? "I have a hypothesis," I said.

Since red and green light combined produce amber, if the LCD is dithered on and off, that is between red and green, the light coming through the LCD should appear to be amber to the human eye, assuming the LCD is dithered faster than retinal neurons can react.

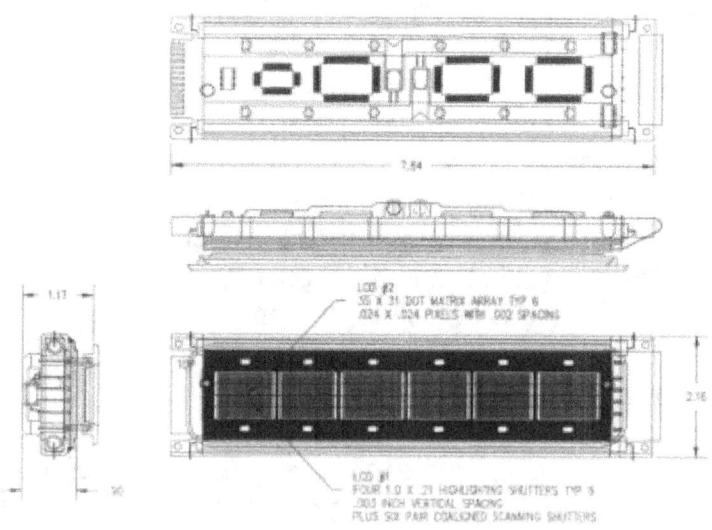

Section 3 – Hypothetical Process of Research

We designed the circuit to dither the LCD rapidly and we got the amber we needed for the control. When we showed the manufacturer of the LCD a model of the control they asked, "How did you do that?"

My answer was, "by getting a very basic understanding of your LCD technology." That's what you have to do to successfully create new products – get a very basic understanding of the need you are addressing and the technology you plan to employ.

I followed the same hypothetical approach when searching for the answer to my two original questions.

Section 4
Dr. Crick's *"Astonishing Hypothesis"*

Dr. Francis Crick, who left his body in 2004, was an avowed atheist. It can be assumed, therefore, that he accepted as fact that the cosmos came into existence from nothing and that man evolved from a lucky protein through natural selection over billions of years. These absurd ideas look like truth in comparison to the statement that the blueprint and biological orders for a complete, living, functioning, thinking human being are stored in the amount of DNA code contained in the zygote (a fertilized ovum). Nothing, however, is further off the mark than Dr. Crick's *Astonishing Hypothesis* (paraphrased: "The human is nothing but bunch of neurons"). My first novel, LINK, refers to Dr. Crick with this statement:

Section 4 – Crick's Astonishing Hypothesis

"What better place to look beyond than through the mind of one who saw to that point but no further."

Dr. Crick apparently believed that there was nothing beyond DNA and that DNA defined the human, all because he did not believe in a Creator.

"A set of blueprints is not a house; the DNA of a zygote is not a human being." Garrett Hardin, Professor of Biology at the University of California at Santa Barbara.

I too doubted had some doubts about the role of a Creator, until I came out of my intellectual hiding place to view Creation with an open mind. Once I accepted His presence, He provided a means for us to communicate. I had many questions and received many answers. The ideas that follow did not come from my mind but did pass through it. Therefore, I want to be certain that any truth found in The Atemporal Particle Theory (hereinafter referred to as the "Theory") is credited to God and any error to me.

On the back cover of Dr. Crick's book, a reviewer says:

"In his book, *The Astonishing Hypothesis*, Nobel laureate Francis Crick boldly straddles the line between science and spirituality by examining the soul from the standpoint of a modern scientist, basing the soul's existence and <u>function</u> on an in-depth examination of how the human brain 'sees.'"

Yet the reader will find that Dr. Crick himself concludes:

Section 4 – Crick's Astonishing Hypothesis

"By the standards of exact science, we do not yet know, even in outline, how our brains produce the vivid visual awareness that we take so much for granted."

In other words, the reviewer says Crick proves his theory based on his understanding of how we all see, yet Crick himself honestly says in his book that scientists have no clue.

It would be good to note here that Jeff Hawkins agrees. He says in his book:

"Our generation has access to a mountain of data about the brain, collected over hundreds of years, and the rate at which they are gathering more data is accelerating. The United States alone has thousands of neuroscientists. Yet we have no productive theories about what intelligence is or how the brain works as a whole."

Hawkins goes on to say, "Crick argued that in spite of a steady accumulation of detailed knowledge about the brain, how the brain works was still a profound mystery. Scientists," Hawkins says, "usually don't write about what they don't know, but Crick didn't care. He was the boy pointing to the emperor with no clothes. [This author thinks Crick was just being honest.] According to Crick," Hawkins continues, "neuroscience was a lot of data without a theory. His exact words were, 'what is conspicuously lacking is a broad framework of ideas.' To me," says Hawkins, "this was the British gentlemen's way of saying, 'We don't have a clue how this thing works.' It was true then, and it's still true today."

Section 4 – Crick's Astonishing Hypothesis

Dr. Crick only feigned his "scientific search for the soul." I propose that the scientific community search for the "soul" (the Particle [17]) in earnest, because therein will lie the answers to visual awareness, consciousness, intelligence and man's search for meaning. Theoretical physicist Stephen William Hawking said, "We pretty well know how the cosmos came into being, but we still don't know why." Scientists must return to considering the "why" in their search for the "how"; and the scientific discovery of the Particle, its purpose and function will lead to unprecedented advances in all biological fields.

---

[17] Particle is defined later.

Section 5
Jeff Hawkins' *On Intelligence*

Jeff Hawkins, the author of *On Intelligence*, doesn't believe in the soul either. He tells us:

"Plato's solution (to understanding how we learn and apply what we've learned) was his famous Theory of Forms. He concluded that our higher minds must be tethered to some transcendent plane of super-reality, where fixed, stable ideas (Forms with a capital F) exist in timeless perfection. Our souls come from this mystical place before birth, he decided, which is where they learned about Forms in the first place. After we're born, we retain latent knowledge of them.  Learning and understanding happen because real-world forms remind us of the Forms to which they correspond. You are able

Section 5 – Hawkins' On Intelligence

to know about circles and dogs because they respectively trigger your memories of Circle and Dog."

Hawkins goes on to say, Plato's solution is "all quite loopy from a modern perspective. But if you strip away the high-flown metaphysics, you can see that he was really talking about 'invariance.' [18] His system of explanation was wildly off the mark, but his intuition that this was one of the most important questions we can ask about our own nature was a bull's-eye."

Hawkins comments about Plato's brilliant solution to how memory and learning must work, give away Hawkins' one failure as a researcher which is his bias against the atemporal. He has thrown the baby out with the bathwater. He missed the point. He closed his eyes to the real solution proposed by Plato.

What if we were to restate Plato's solution "from a modern perspective," which is what an open-minded researcher would have done. It might read something like this:

"Plato concluded that our higher minds (the neurons of the neocortex) must be tethered to (i.e., able to communicate with) some transcendent plane of super-reality (an atemporal Particle of zero mass), where

---

[18] "invariance" as used by Hawkins means an invariant representation, stored in the cortex, of an image, sound, smell or touch which is recalled and compared to the real world experience providing immediate recognition of what is being experienced.

27

Section 5 – Hawkins' On Intelligence

fixed, stable ideas (ancestral memories [19]) exist in timeless perfection (i.e., stored atemporally and perfectly). After we're born, we retain latent knowledge of them (the ancestral memories)."

"I'm an atheist," Hawkins said in an interview. "I'm not militant about it or anything like that, but I don't think you need religion to be a good guy, to be a smart guy, to do good or to be kind and caring or anything like that."

He's right. You can be a good, smart, kind and caring guy and not believe in God. My worst best friend is a brilliant designer, potter, musician and a kind, caring guy who even sings in his wife's church choir, and yet he says he doesn't believe in God. Unfortunately, because Hawkins does not believe in a Creator, he has trouble believing in the atemporal part of reality.

In a discussion of the possibility that we live in an essentially atemporal universe, an unidentified scientist [20] said:

"There is an intuitive feeling that if we do live in a universe that lacks fundamental temporality, (this concept) might provide answers to some of the most difficult questions in quantum physics[ii] and relativity."

Now that's a scientist open to possibilities without bias!

---

[19] Discussed later.
[20] Could have been any one of a number of scientists discussing this question.

Section 5 – Hawkins' On Intelligence

Plato was on the right track and so is Hawkins. Hawkins has developed a brilliant theory of how the brain produces intelligence (not consciousness). I strongly recommend his book, *On Intelligence*. Like Crick, Hawkins just can't get past the physical aspect of reality, because of his atheistic bias. What a shame.

Section 6
The Foundational Concept

Search for a basic understanding

We must start our discussion with a very basic understanding of our existence. If we don't have a grasp on who we are, then what is the purpose?

I knew from the beginning of my study, while I was developing the Theory, that I did not have a very basic understanding of the primary factors (like I had to with the touch-screen LCD project). As a scientist, that made me very uncomfortable. I knew there were two states of existence, the state in which you and I currently exist, a state in which we are limited by time; and another state in which a God, Creator, Supreme Being ("God") must exist, not limited by time. The God of the Bible said (yes, I will even quote the Bible from time to time - Douay–Rheims version [21]), "I am who am," [22] meaning He just is. Here is the Hebrew text from the Torah:

---

[21] 1899 American Edition
[22] Exodus 3:14

Section 6 – The Foundational Concept

יד וַיֹּאמֶר אֱלֹהִים אֶל-מֹשֶׁה, אֶהְיֶה אֲשֶׁר אֶהְיֶה; וַיֹּאמֶר, כֹּה תֹאמַר לִבְנֵי יִשְׂרָאֵל, אֶהְיֶה, שְׁלָחַנִי אֲלֵיכֶם.

It reads from right to left. This is the subject phrase:

אֶהְיֶה אֲשֶׁר אֶהְיֶה

It translates literally, "Am who am." I am who is. God just is.[iii]

However, there seemed to be, based on Christian, 2,000-year-old beliefs (which we cannot ignore), that there is another state between the state in which we exist, i.e., the corporeal state and, as Christians say, the spiritual realm, an incorporeal [23] state. But what are the characteristics and limitations of the incorporeal state? Nowhere was I able to find and read a clear explanation, not in any Bible translation or any other source.

Sequence

One night I was driving home from a design engineering consulting assignment, quite dead tired, able to think of nothing but relaxing at home with my wife, Kim. I came to a stop light where the road ended at a highway. I can still see in my mind the red light I was watching when then quite suddenly, I had the answer to my question. We, in a state I now call the **Temporal State**, are not just limited by time. We are also limited by the sequence of events. That had never occurred to me.

---

[23] not composed of matter; having no material existence. Source: Oxford Dictionary

Section 6 – The Foundational Concept

God cannot be limited by time nor the sequence of events. There must be no limitations in a God. There is a third state, I came to understand, which I now call the **Sequential State**. Beings in the Sequential State are limited only by sequence, not (necessarily) by time. I drove home and told Kim what I had learned. She said, "You must write that down." I told her I would, but I didn't have to. I will never forget the moment I learned this new truth. It changed everything.

Temporal and Sequential States

What does it mean for us in the Temporal State to be limited by time and sequence? Think of it this way: you must open a closed door before you enter a room; that takes time. You must travel to your relative's home to hear what they have to say; travel takes time. You must open your car door and sit in the driver's seat before starting the car. That takes time. All this sounds obvious, I know. Well, nonetheless, that describes the Temporal State. Now, if you, as a physical being, could function in the spiritual, Sequential State, **once you will the action**: You can instantly be in a room having passed through the closed door; you can instantly be in the home of your relatives to hear what they have to say; you can instantly be in the driver's seat of your car ready to start your car. However, in the Sequential State, you cannot undo having passed through the closed door and entered a room; you cannot undo having arrived at the home of your relatives; you cannot undo having instantly found yourself in the driver's seat of your car.

Section 6 – The Foundational Concept

There, you see, is the limitation of sequence. Now, it is hard to imagine taking no time to do things but, and here is some shocking news, two component parts of you are actually in the Sequential State.

Concurrent State

To stretch your thought process even further, consider the **Concurrent State**, a state in which only GOD exists, unlimited by time and unlimited by sequence. Quoting again from the Bible, Jesus, who said He is God, gave us a perfect example of this when He said to the Jews of his time, "Your father Abraham rejoiced at the thought of seeing my day; he saw it and was glad."
"You are not yet fifty years old," they said to him, "and you have seen Abraham!" "Very truly I tell you," Jesus answered, "before Abraham was born, I am!" [24] Yet Jesus was born of the virgin Mary just 30 years before he said that. And, according to the Gospel of the Apostle John, He existed before the cosmos came into being, which He Himself created.[25]

So, let me review this foundational part of the Theory behind this new science:

Three states of existence:
1. The Temporal State in which beings are limited by time and sequence,
2. The Sequential State in which beings are limited only by sequence, and

---

[24] John 8:56-58
[25] John 1:1-5

Section 6 – The Foundational Concept

3. The Concurrent State in which God is not limited by time nor Sequence.

You are presently experiencing the Temporal State in which each sequential event is spaced by time. The cause of something happening always precedes that which happens. Events that occur cannot "unoccur" (try looking that up in the dictionary). There is no going back in time nor in sequence. In the Sequential State things can happen instantly, with no time between events, but there is still no going back in the sequence of events.

Important: You must accept the hypothesis that there are three states of existence, as defined above, in order to move forward with the Atemporal Particle Theory. If you can't accept the hypothesis, at least for the discussion that follows, then you may as well close the book now.

Section 7
Creation[26]

God the Father, God the Son and God the Holy Spirit have no limitations whatsoever. God is not limited by time nor the sequence of events. Jesus could place a mountain in the middle of Montana, next week, that is subsequently discovered to have been there for 250 million years. Jesus can accurately say, "Before Abraham came to be, I am,"[27] yet He was born of Mary approximately 2,020 years ago.

"I am" (see End Note [iv]) is the same perfect description God (Jesus) gave Moses when Moses asked, "If I shall go to the children of Israel, and say to them, 'The God of your fathers has sent me to you.' If they should say to me, 'What is His name?' What shall I say to them?" He said to Moses: "I am who am. Thus, shall you say

---

[26] The entire Creation Section was taken from the booklet, *Jesus – a Scientist's Perspective*, by John Beiswenger
[27] John 8:58

Section 7 – Creation

to the children of Israel: He who is, has sent me to you."[28] Keep these things in mind as we discuss Creation.

"All things came into being through Him (Jesus), and apart from Him nothing came into being that has come into being. In Him was life, and the life was the Light of men."[29] Every-thing had a beginning with Jesus, not a "pre-incarnate" Jesus nor a "Christophany" like some have to say; those who cannot accept that God is not limited by time nor the sequence of events.

Then God (Jesus) said, "Let there be light"; and there was light."[30] "The universe, and time itself, had a beginning in the Big Bang, about 15 billion years ago."[31] Most scientists agree that the universe was created, as Hawking said, beginning with an infinitesimally small, infinitely hot, infinitely dense **Cosmic Singularity**[32], which had to explode the instant it was created. Everything that came into being was therewith created by Jesus. He personally created every Black Hole[33] from which every galaxy developed and therewith every star, every planet, every moon and even every asteroid and comet, all from the Cosmic Singularity **which Jesus created from nothing**. He did what we believe should have, and appears to have,

---

[28] Exodus 3:13-15
[29] John 1:3-4
[30] Genesis 1:3
[31] Stephen Hawking
[32] Containing all the matter and energy of the Universe.
[33] Black holes contain a spacetime singularity at their center.

Section 7 – Creation

taken billions of years, all in the time-period of a single earth day.[v]

Then Jesus prepared the earth, a special planet, for what was to come. He created the atmosphere surrounding the earth, separating the water on the earth from the clouds above it, and then He said, "Let there be an expanse in the midst of the waters, and let it separate the waters from the waters."[34] He prepared the earth in just 24 of our earth's hours.

Next (if that is an appropriate term for the actions of God), Jesus created the geology of the earth, with mountains, valleys, seas, oceans, lakes and rivers, all of His own design. What will be determined to have taken many, many millions of years, Jesus caused to happen in a matter of hours. After the earth's geology was created, He created a **Botanical Singularity** with every form of meristematic[35] cell capable of growing into every plant type on the earth. "And He said: Let the earth bring forth the green herb, and such as may seed, and the fruit tree yielding fruit after its kind, which may have seed in itself, upon the earth. And it was so done."[36] With this Jesus also personally designed and created every living plant and tree (down to every leaf), flower, fruit, and vegetable. In 24 hours of our time, Jesus created the geology of the earth and then He created all plant life from the Botanical Singularity **which Jesus created from nothing**.

---

[34] Genesis 1:6
[35] embryonic tissue in plants; undifferentiated, growing, actively dividing cells.
[36] Genesis 1:11

Section 7 - Creation

Jesus then gave order to the universe, created during the Big Bang, including each solar system, so that the planets, especially the earth had a perfect source of heat and light. This happened in another 24 of our hours. From the earth we can look into the sky and see His amazing, creative work.

It was then that Jesus created the **Biological Singularity** containing the stem cells, both corporeal and incorporeal, of every creature having life, from the simplest life form to the most complex, each with a controlling, atemporal component ("Particle") at its functional center.

And "God (Jesus) said: Let the waters bring forth the creeping creature having life, and the fowl that may fly over the earth under the firmament of heaven. And God created the great whales, and every living and moving creature, which the waters brought forth, according to their kinds, and every winged fowl according to its kind. And He saw that it was good. And He blessed them, saying: Increase and multiply, and fill the waters of the sea: and let the birds be multiplied upon the earth."[37] "And God said: Let the earth bring forth the living creature in its kind, cattle and creeping things, and beasts of the earth, according to their kinds. And it was so done."[38] And with that, Jesus personally designed and created all living creatures, every individual species, all created in 24 earth-hours from the Biological Singularity **which Jesus created from nothing**.

---

[37] Genesis 1:20-22
[38] Genesis 1:24

Section 7 - Creation

"And He said: Let us make man to our image and likeness: and let him have dominion over the fishes of the sea, and the fowls of the air, and the beasts, and the whole earth, and every creeping creature that moves upon the earth."[39] Jesus, alive with body and spirit, created the cosmos and everything and everyone in it, and it was Jesus who said, "Let us make man in our image (spirit) and likeness (body)."[40] Who else in the Holy Trinity do we resemble?

---

[39] Genesis 1:26
[40] taken from Genesis 1:26

Section 8
The Atemporal Particle Theory

The Atemporal Particle Theory stated concisely:

Coincident [41] with every human, living, corporeal cell is an incorporeal cell, and at the functional center of this coexistence [42] is an atemporal Particle, which directs all functions of the cells, stores all memory and co-generates consciousness.

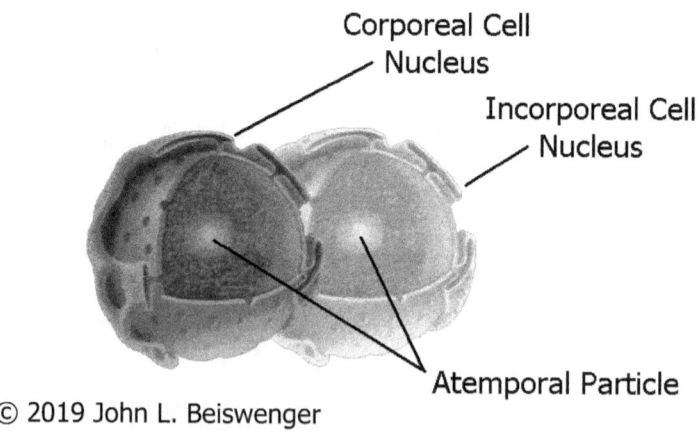

© 2019 John L. Beiswenger

---

[41] Occupying the same space or time (Merriam-Webster)
[42] Noun - Entities existing together (Author)

Section 8 – The Atemporal Particle Theory

I will unpack the Theory for you starting with:

"Coincident with every human, living, corporeal cell is an incorporeal cell."

The human body is comprised of trillions of these coexisting cells, consisting of hundreds of cell types. The resulting, complete corporeal body and the complete incorporeal body are therefore also coincident; they occupy the same space, at the same time.[43] The corporeal body and incorporeal body are bound together (defining "Unity" [44]) with every movement made. The Unity is inseparable, except as later explained.

". . . at the functional center of this coexistence is an atemporal Particle."

The same Particle is at the functional center of every cell in the corporeal body and incorporeal body; possible because it is atemporal, has zero mass and therefore can be at the functional center of each cell simultaneously – a true, biological singularity.

"(The Particle) . . . directs all functions of the cells."

It is the Particle which directs the biological functions of all cells, with instructions which are filtered, figuratively, through the hereditary traits encoded by the DNA in

---

[43] Except, of course, when errors in DNA produce significant differences in structure.
[44] The Unity is one person.

Section 8 – The Atemporal Particle Theory

the nucleus of the cells. It is also the Particle which generates an electromagnetic field in the nuclei of neurons causing action potentials to occur.[vi] Without the Particle, the human would not be alive (exist).

"(The Particle) . . . stores all memory."

It is the Particle that is the repository of all memories and those of the individual's ancestors, not neurons, neural networks, synapses and no, not DNA. Stored in an individual's Particle is the memory of everything he/she has ever experienced. Memories are returned to the same neural network which placed them there in the first place.[45] If that neural network is damaged or missing, the memory may not be recallable in part or at all.[46]

"(The Particle) . . . co-generates consciousness."

The Particle is at the functional center of every one of the billions of neurons throughout the corporeal/incorporeal body Unity [47]. The Particle is communicating continuously with every cell. With the cells in the central nervous system (neurons), there are two way communications.

---

[45] Which is why an individual, in the physical, cannot readily recall ancestral memories.

[46] Note: There can be no damaged or missing neural networks in the Sequential State brain and therefore all memory can be recalled by the Sequential State brain.

[47] It would be a mistake to think of the corporeal body and incorporeal body as two separate entities. They comprise one person.

Section 8 – The Atemporal Particle Theory

Sensory neurons, such as retinal neurons, transfer information at random to the brain. Cortical neurons in turn communicate this incomprehensible information constantly with the Particle. On a computer screen it would appear as noise. The Particle stores and instantly returns the information received as pulsed, comprehensible percepts, making consciousness possible.[vii]

That's the Theory. Now let's look at the evidence.

Section 9
Evidence of the Incorporeal Body

As a research engineer, I learned not to believe anything I hear, see or read. Better stated, I learned to confirm the veracity of everything I needed to believe. So, where is the evidence for the Atemporal Particle Theory? I will herewith present support for every aspect of the Theory by showing how the Theory explains the following phenomena.

Near-death experiences

The incorporeal body possessed by all living humans is well evidenced in near-death experiences during which the incorporeal body has clearly separated from the corporeal body. The incorporeal body is generally invisible to those in the Temporal State, it can observe what is going on around it, can hear, see and later recall the activity; it can pass through physical objects and move through great distances at will. This has been experienced by thousands.

Section 9 – Evidence of the Incorporeal Body

Here is why/how it occurs: If for any reason, the major control centers in the corporeal brain stop communicating with the Particle, the incorporeal body may separate from the corporeal body. This happens when the corporeal brain can no longer function or when, through sensory deprivation, the corporeal brain is not transmitting information to the Particle and an out-of-body-experience occurs.

Near-death-experiences are well documented. Dr. Michael Sabom is a cardiologist who detailed the near-death experience of nearly 50 individuals. One was a woman he named Pam Reynolds. She underwent a rare operation to remove a giant aneurysm in her brain that threatened her life. The operation required that Pam's body temperature be lowered to 60 degrees, her heartbeat and breathing stopped and the blood drained from her head. She was clinically quite dead. During that time, she had an out-of-body experience which was later verified to be true. Here's her entire statement from Dr. Sabom's book, "Light & Death."

Pam Reynolds

"I remember seeing several things in the operating room when I was looking down (from above the operating table). The saw (the surgeon was using) looked like an electric toothbrush. And the saw had interchangeable blades too, but these blades were in what looked like a socket wrench case.

Section 9 – Evidence of the Incorporeal Body

Someone said something about my veins and arteries being very small. I believe it was a female voice (speaking) and that it was Dr. Murray, the cardiologist. I remember thinking that I should have told her about that.

There was a sensation like being pulled, but not against (my) will. I was going on my own accord, because I wanted to go. It was like being taken up in a tornado vortex, only you're not spinning around. At the end there was this little tiny pinpoint of light that kept getting bigger and bigger. The light was incredibly bright. I noticed that as I began to discern different figures in the light, they began to form shapes I could recognize and understand.

I could see that one of them was my grandmother. Everyone I saw, looking back on it, fit perfectly into my understanding of what that person looked like at their best during their lives. I recognized a lot of people. My uncle Gene was there; so was my great-great aunt Maggie. On (my father's) side of the family, my grandfather was there. They were specifically taking care of me, looking after me. They would not permit me to go further.

It was communicated to me that if I went all the way into the light, something would happen to me corporeally. They would be unable to put this _me_ back into the body _me_. I wanted to go into

Section 9 – Evidence of the Incorporeal Body

the light, but I also wanted to come back. I had children to be reared.

My uncle . . . took me back through the end of the tunnel. Everything was fine. I did want to go. But then I got to the end of (the tunnel) and saw the thing, my body. I didn't want to get into it. It looked like what it was, dead. It scared me and I didn't want to look at it. I felt a definite repelling and at the same time pulling from the body. The body was pulling, and the tunnel was pushing. It was like diving into a pool of ice water . . . It hurt.

When I came back (into my body), they were playing 'Hotel California' (in the operating room) and the line was, 'You can check out anytime, but you can never leave.' When I regained consciousness, I mentioned to Dr. Brown that (playing that song) was incredibly insensitive."

Savant Abilities

Communications between the neurons of the corporeal body's cells and the Particle are slowed by the physical limitations of the corporeal body's brain especially as the body ages. On the other hand, communications between the neurons of the incorporeal body and the Particle are not slowed by any physical limitations. Time is not a limitation in this state.

If a part of the corporeal body's brain is diseased, undeveloped or removed, the neurons of the incorporeal

Section 9 – Evidence of the Incorporeal Body

brain may function in its place, and the results may be far in excess of what would be considered normal.

This is true for savants whose corporeal brains were damaged[viii] or never developed properly. Stephen Wiltshire, for example, was autistic and mute when he started drawing. His extraordinary abilities include drawing extremely accurate panoramic pictures of landscapes after seeing them for a short time during a helicopter ride. His incorporeal brain, functioning in place of the undeveloped portion of his corporeal brain, has no trouble recalling everything he has seen in great detail.

Leslie Lemke has severe birth defects. Blind from birth, he had to be force fed to learn how to swallow, and only very gradually learned to walk. Suddenly at age 14 he was found playing Tchaikovsky's Piano Concerto No.1, with no previous musical training, which he heard once on television. His incorporeal brain, functioning in place of the undeveloped portion of his corporeal brain, is able to recall fully and accurately everything he has heard. It is not surprising then, that if and when the artist he is listening to makes a minor mistake, Leslie makes the same mistake.

The same is true for this savant. After a head injury, Alonzo Clemons, has had severe difficulties in functioning, unable to feed himself, but he can accurately sculpt in three dimensions anything he briefly sees once.

Section 9 – Evidence of the Incorporeal Body

Born prematurely, Derek remained in the hospital for three months and technically "died" several times before he was finally strong enough to go home. Left blind and with severe cognitive impairment, he taught himself to play the piano and can perform any music upon hearing it once. His incorporeal brain, functioning in place of the undeveloped portion of his corporeal brain has no difficulty recalling everything he has heard with perfect accuracy.

Dr. Oliver Wolf Sacks, a British neurologist describes his experience with twins known as John and Michael.[48]

"I first met the twins in 1966 in a state hospital; they were already well known. My own first sight of the 'natural' powers, and 'natural' mode of the twins, came in a similar, spontaneous, and (I could not help feeling) rather comic manner. A box of matches on their table fell and discharged its contents on the floor: '111,' they both cried simultaneously; and then, in a murmur, John said '37'. Michael repeated this, John said it a third time and (they) stopped. I counted the matches - it took me some time - and there were 111." Their incorporeal brains, functioning in place of the undeveloped portion of their corporeal brains, saw and instantly counted all of the matches.

There are hundreds of such examples.

---

[48] Excerpt from Chapter 23 of Oliver Sacks' The Man Who Mistook His Wife For a Hat.

Section 9 – Evidence of the Incorporeal Body

Hyperthymesia Syndrome

Savant-like abilities occur rarely in relatively normal people. Television star Marilu Henner is one of a half-dozen known individuals with highly superior autobiographical memory, medically called Hyperthymesia Syndrome. She can recall in detail every event which happened in her life including the date and day of the week on which it happened. She appears able to recall memories from her Particle through the use of her incorporeal brain.

Phantom Limb Phenomena

Case 4, in the same publication referenced above (see footnote 9 on Page 14), a woman, now age 31, was born with a shortened and deformed right leg and underwent amputation of the foot and deformed part of the leg at age 3. Beginning at age 6, she developed a full-length phantom leg and foot which touched the floor.

If a corporeal body's limb is removed, the incorporeal body's limb will still be there, which explains the phantom-limb phenomena.

"In 2012 V.S. Ramachandran and Paul McGeoch [49] reported the case of a 57-year-old woman (known as R.N.) who was born with a deformed right hand consisting of only three fingers and a rudimentary

---

[49] McGeoch, P., and Ramachandran, V., (2012), *The appearance of new phantom fingers post-amputation in a phocomelus, Neurocase*, 18 (2), 95-97.

Section 9 — Evidence of the Incorporeal Body

thumb. After a car crash at the age of 18, the woman's deformed hand was amputated, which gave rise to feelings of a phantom hand. The phantom hand was experienced, however, as having all five fingers (although some of the digits were foreshortened). 35 years after her accident, the woman was referred for treatment after her phantom hand had become unbearably painful. McGeoch and Ramachandran, a trained R.N., used a mirror-box visual feedback, for 30 minutes a day, in which the reflection of the woman's healthy left-hand was seen as superimposed onto where she felt her phantom right hand to be. After two weeks she was able to move her phantom fingers and was relieved of pain. Coincidently, she also experienced that all five of her phantom fingers were now normal length. Ramachandran and McGeoch stated that this case provides evidence that the brain has an innate (hard-wired) template of a fully formed hand."

It is well beyond reason to believe that the cerebral cortex , "about 2 millimeters thick, stretched flat, roughly the size of a large dinner napkin" described again in a later Section, Memory and Consciousness, contains the "template" of every one of the trillions of cells in the body.

I meet weekly with a physician as part of the work I do. He told me he too had a patient who had lost her left forearm, wrist and hand. She complained to him that her left phantom arm had made a tight fist and it was somehow painful to her. He set up a mirror, as explained above, in which she could see what looked like her left arm (just a reflection of her right arm) and he

Section 9 – Evidence of the Incorporeal Body

told her to relax both fists. She did, her phantom fist relaxed, and the pain went away. Some explain this by saying the missing arm was in her memory. Well, a girl who was born without an arm, was also aware of her incorporeal body's arm, and she clearly had no memory of it. (repeated earlier)

Here's another report:

The Consciousness of Lost Limbs, By William James (1887) First published in Proceedings of the American Society for Psychical Research, 1, 249-258. Posted, March 2003:

> "I have obtained first-hand information from a hundred and eighty-five amputated persons. Generally, the position of the (phantom) lost leg follows that of the stump and artificial leg. If one is flexed the other seems flexed; if one is extended so is the other; if one swings in walking the other swings with it.
>
> About a hundred of the cases who feel their feet, affirm that they can "work" or "wiggle" their toes at will. Almost always when the will is exerted to move the toes, actual contraction may be perceived in the muscles of the stump. One might, therefore, expect that where the toe-moving muscles were cut off, the sense of the toes being moved might disappear. But this is not the case."

Section 9 – Evidence of the Incorporeal Body

Hemispherectomy

As we said earlier, if part of the corporeal brain is removed, inactivated or damaged, the incorporeal brain may supplement that part which is not functioning. The entire left side of a child's brain was removed to stop her seizures. Though she still requires extensive care, her parents report that Kara is doing well physically and emotionally and is looking forward to preschool.

Dr. Ben Carson also removed the entire left brain of a child named Miranda, with unexpected results. Here is Dr. Carson's report:

> "Dr. Neville (Knuckey) and I, and the rest of our team, knew we had successfully removed the left hemisphere of Maranda's brain. We didn't know if her seizures would stop. We didn't know if Maranda would ever walk or talk again. We could only do one thing - wait and see. The (parents), alert to every sound, heard the gurney creaking down the hallway and ran to meet us.
> "Wait!" (her mother) called softly. She went to the gurney bent down and kissed her daughter. Maranda's eyes fluttered open for a second. "I love you, Mommy and Daddy," she said. I just stood there, amazed and excited, as I silently shared in that incredible moment.
>
> We had hoped for recovery, but none of us had considered that she could be so alert so quickly. Maranda had opened her eyes. She recognized

Section 9 – Evidence of the Incorporeal Body

her parents. She was talking, hearing, thinking, responding.

We had removed the left half of her brain, the dominant part that controls the speech area. Yet Maranda was talking!

She was a little restless, uncomfortable on the narrow gurney, and stretched her right leg, moved her right arm - the side controlled by the half of her brain we had removed!" "Gifted Hands," 20th Anniversary Edition: The Ben Carson Story [Kindle Edition] published in 2011.

I believe that Maranda's left incorporeal brain hemisphere was functioning in place of the removed corporeal hemisphere; true also for Kara.

Repetitive Transcranial magnetic stimulation

Dr. Allan Snyder, head of the Centre for the Mind at the University of Sydney, uses transcranial magnetic stimulation to temporarily shut down the left hemisphere of the brain, where speech and short-term memory are supported, inducing surprising savant-like skills in healthy people. Here's his report, referenced earlier:

> "To test the suggestion that normal individuals have the capacity for savant numerosity, we temporarily simulated the savant condition in normal people by inhibiting the left anterior temporal lobe of twelve participants with

Section 9 – Evidence of the Incorporeal Body

repetitive transcranial magnetic stimulation (rTMS). This (brain) site has been implicated in the savant condition. Ten participants improved their ability to accurately guess the number of discrete items immediately following rTMS and, of these, eight became worse at guessing as the effects of the pulses receded. The probability of as many as eight out of twelve people doing best just after rTMS and not after sham stimulation by chance alone is less than one in one thousand."

When the Particle cannot communicate with the neurons in the left anterior temporal lobe of the participants, their corresponding incorporeal brain segment functions in its place. The incorporeal brain is not limited by time and can quickly solve the problems. In Dr. Snyder's full report, he refers to Dr. Oliver Wolf Sacks report regarding the twins known as John and Michael.

Frontotemporal dementia

It also happens through disease. Dr. Bruce Miller, Professor of Neurology at the University of California, has seen patients, with frontotemporal dementia, spontaneously develop both interest and skill in art and music. Dr. Miller has also seen physiological similarities in the brains of autistic savants. The incorporeal brain supplements the diseased section. Here's the report:

> "Dr. Miller, working with patients having frontotemporal dementia, a degenerative brain disease that strikes people in their fifties and sixties, has seen these patients spontaneously

Section 9 – Evidence of the Incorporeal Body

develop both interest and skill in art and music. Patients with damage in this area can't name what they're looking at, but they can often paint it beautifully.

Dr. Miller has also seen physiological similarities in the brains of autistic savants and patients with frontotemporal dementia. When he performed brain-imaging studies on an autistic savant artist who started drawing horses at 18 months, he saw abnormalities similar to those of artists with frontotemporal dementia: decreased blood flow (therefore oxygen) and slowed neuronal firing in the left temporal lobe."

The Autistic Twins

There's more to learn about the Atemporal Particle Theory from the Twins report by Dr. Oliver Wolf Sacks.

When they seemed to be able to count in an instant the 111 matches that fell to the floor, Dr. Sacks asked them, "How could you count the matches so quickly?" "We didn't count them," they said, "We saw the 111."

Let's try that with you. In your mind, without speaking the numbers, multiply 12 times 12. You have the answer in your mind. You see 144, just like the twins did, only you did the calculation. Where did the 111 they saw come from? How were the matches counted instantly, because they had to have been counted?

The answer is the twins actually did count the matches.

Section 9 – Evidence of the Incorporeal Body

Their incorporeal eyes and brain saw them and counted the matches instantly. The twin's corporal brains were unaware of this happening, focused like ours on the physical world around us, but not unlike intuition-delivered information, they saw the number 111, the final count number.

The twins were brain damaged from birth and have an IQ of 60, yet they have memories that go back accurately to when they were 4-years old. They were unable to complete the simplest arithmetic problem. Consider this:

"The next time I saw them," said Dr. Sacks, paraphrased from his earlier referenced book, "they seemed to be locked in a singular, purely numerical conversation. John would say a six-figure number, Michael would nod and smile, and would throw out a new six-figure number. Had the numbers any meaning, I wondered? I noted down the numbers. As soon as I got home, I pulled out my tables of powers, factors, logarithms and primes. I had a hunch and now I confirmed it. All of the numbers, the six figure numbers, which the twins had exchanged were prime numbers, numbers that can be evenly divided by no other whole number than itself or (the number) one."

The twins corporeal brains could not do what I asked you to do in your mind, i.e., 12 x 12 = 144, but their incorporeal brains could mentally search for a 6-digit number that was prime, and the twins corporeal brains could "see" the resulting numbers, just like you could see the 144.

Section 9 – Evidence of the Incorporeal Body

Blindsightedness

"Blindsight is the ability of people who are cortically blind due to damage to their primary visual cortex to respond to visual stimuli that they do not physically see." (Wikipedia – "Blindsight")

"Of particular interest has been the fact that they can sense emotion: when presented with faces, they can tell whether it is happy or sad, angry or surprised, and they even start to unconsciously mimic the expressions. Even though they did not report anything at a conscious level, we could show a change in attitude, a synchronization of emotional expressions to the pictures in their blind field. Besides mirroring expressions, they also show physiological signs of stress when they see a picture of a frightened face." (By David Robson, BBC, "The strangest form of consciousness")

It is the Sequential State body which sees the visual stimuli relating the stimuli (and even emotion) to the Temporal State brain through intuition.[50]

---

[50] See Page 97, Intuition Sub-Section.

Section 10
Evidence of the Atemporal Particle

The Binding Problem

The Binding Problem, which obsessed Dr. Francis Crick and many other scientists,[51] is defined by him as the problem of how (a set of) neurons temporarily become active as a unit, is explained by the Atemporal Particle Theory.

Since the Particle is at the functional center of every neuron, and since the Particle is processing all of the information sent to the brains, it is the Particle which communicates ("binds") the information with "all those neurons actively responding to different aspects of a perceived object."

The Atemporal Particle Theory seems to be the only theory which proposes an answer to the Binding Problem.

---

[51] See End Note [i]

Section 10 – Evidence of the Atemporal Particle

Cellular Control

In the book *Molecular Biology of the Cell, 4th edition,*[52] referenced also earlier, the authors compare the cellular-control process to that of an "automatic clothes-washing machine. The washing machine functions through a series of stages: it takes in water, mixes it with detergent, washes the clothes, rinses them, and spins them dry. These essential processes of the wash cycle are analogous to the essential processes of the cell cycle - DNA replication, mitosis, and so on. In both cases, a <u>central controller</u> triggers each process in a set sequence." And they add, "Sensors, for example, detect the completion of DNA synthesis (or the successful filling of the washtub), and, if some malfunction prevents the successful completion of this process, signals are sent to the control system to delay progression to the next phase."

In addition to being a scientist, I am also a design engineer and I designed a surgical instrument sterilizer which used a toxic chemical. Like the washing machine they referenced, the sterilizer also took in water, mixed it with the sterilizing compound, sterilized the instruments, and rinsed them thoroughly of any toxic material. If something went wrong during the cycle, the sterilizer immediate stopped the process and flushed away the toxic chemical. A <u>central controller</u> triggers each process in a set sequence. And what controls the central controller?

---

[52] Molecular Biology of the Cell, 4th edition, Bruce Alberts, Alexander Johnson, Julian Lewis, Martin Raff, Keith Roberts, and Peter Walter. New York: Garland Science; 2002.

Section 10 – Evidence of the Atemporal Particle

The design engineers of the wash machine they referenced, and the design engineer of the above surgical instrument sterilizer (me), indirectly control the central controller with the software written to direct and monitor each phase of the wash cycle, <u>started by a human pressing a momentary switch</u>. If it is the cyclical changes in cyclin levels which result in the cyclic assembly and activation of the cyclin-Cdk complexes, and that activation in turn triggers cell-cycle events, then there must be an information and control source involved, such as the Particle.

Scientists must be aware of the need for a heretofore unknown source of information within each cell.[ix]

Consider reading these two books:

> ***Wetware, A computer in every living cell*** by Dennis Bray, the scientist who believes the components of every cell form a micro-computer

Section 10 – Evidence of the Atemporal Particle

within each cell which has been programmed by . . . evolution. (Right)

In the beginning was information by Dr. Werner Gitt., a scientist (who) explains the incredible design in nature. Dr. Gitt is a Christian. His book is very technical, but a worthwhile read.

Amputation and Regrowth

"Kids will actually regrow a pretty good fingertip, after amputation, if you just leave it alone," says Dr. Christopher Allan, from the University of Washington Medicine Hand Center. The orthopedic surgeon saw this out a few years ago when an 8-year-old girl stuck her finger into the spokes of her brother's bike. The wheel sliced off her middle finger, near the nail cuticle, and her parents rushed to the ER to have it sewn back on. Allan specializes in hand reconstruction, but he couldn't find the tiny artery he needed to reconnect. So he opted instead for what surgeons call a 'biological dressing.' "Just stick the tip back on and hope for the best," he says. "The girl came back in a few weeks with the old fingertip in a bag and a new one on her hand. It was far better than anything that I could have given her with a graft or surgery."

The Particle supplied the information to grow back the child's finger as it does to heal any wounds.

Section 11
Memory and Consciousness

"There is considerable evidence that our retention is much better than our normal recall would lead us to expect - indeed we may retain all of our experience." Dr. Robert C. Gilman, Ph.D., Astrophysicist.

Dr. Paul Pearsall, author of *The Heart's Code*, is a psycho-neuro-immunologist, a licensed psychologist who studies the relationship between the brain, immune system, and external factors. He has documented dozens of cases in which heart transplant recipients have received some of the memories of the donors.[x] For example, a little girl who received the heart of a murdered child was reported to have 'recalled' the child's killer so well that she described him, and he was eventually convicted, as described earlier.

Dr. Pearsall believes cells story memory. "That every cell seems to be able to somehow "remember" what it is supposed to do and when, where and with which other cells it is supposed to do it, is one of the still unexplained miracles of life." The Particle remains with the physical body (and body parts) until corruption takes place,

Section 11 – Memory and Consciousness

explaining this "miracle of life."

Yes, I think you should be aware of this when advising people about receiving a transplant organ, but memories are not stored in the cells themselves as Dr. Pearsall suggests. They are stored in the Particle at the functional center of every cell.

In his paper entitled, *Are Memories Really Stored in the Brain?* Nicholas H.E. Prince, Mathematical Physicist writes, "Essentially the thesis outlined in this paper begins at the outset by assuming that the brain itself does not store (long term) memories at all, but rather retrieves them from an external store. Indeed, the implications of such a mechanism, if real, would be far reaching." He's right. He concludes, "Memories are recovered atemporally (from a timeless state)," but he does not explain how.

[See also the report in the NAMAH Journal in End Note[xi]]

Ray Tillis, M.D., Professor of Geriatric Medicine, UK, wrote a fascinating article in the *New Scientist* magazine about memory: "Memory is typically (viewed) as being "stored" (in the brain). But when I 'remember,' I explicitly reach out of the present to something that is explicitly past. (The brain is a) physical structure (knowing only the) present state. In other words, the sense of the past cannot exist in a physical (temporal) system."

Temporal-State data communicated to the Particle is stored as Sequential-State data (defining "Parallata," pl.

Section 11 – Memory and Consciousness

of Parallatum). Memories are a series of Parallata stored sequentially and permanently in the Particle.

The Particle is able to recall memories instantly. When a memory is recalled, it is returned to the same neural network that transferred the Parallata where it is "re-experienced."

> **How often have you heard a memory recalled with, "I can almost feel the sun on my face and smell the spring air"?** [xii]

If the neural network that transferred the stored data to the Particle has been damaged or destroyed, and not replaced, the memory cannot, under normal conditions, be recalled, completely, accurately or at all.

Most scientists believe that if you damage a certain area of the neocortex, memories stored in that area can be lost, which proves to their satisfaction that memories are actually stored in the brain. The above hypothetical statements suggest that memories recalled from the Particle are re-experienced through the same neural-network that stored them. Therefore, if those networks are damaged, the memories cannot be recalled or re-experienced.

Hawkins teaches, "Even before neuroscientists were able to discern anything helpful about the circuitry of the cortex, they knew some mental functions were localized to certain regions of it. If a stroke knocks out Joe's right parietal lobe, he can lose his ability to perceive – or even conceive of – anything on the left side of his body or in

Section 11 – Memory and Consciousness

the left half of space around himself. A stroke in the left frontal region known as Broca's area, by contrast, compromises his ability to use the rules of grammar although his vocabulary and his ability to understand the meaning of words are unchanged. A stroke in an area called the fusiform gyrus can knock out the ability to recognize faces – Joe can't recognize his mother, his children or even his own face in a photograph. Deeply fascinating disorders like these gave early neuroscientists the notion that the cortex consists of many functional regions."

The Theory suggests, it is the transfer of information from the brain to the Particle that permits the "stacking" of sequential data in the Particle, and it is the pulsed response of the Particle that introduces time between the synchronous Parallata returning to the neurons of the brain. If it were not for the cooperative process of the brain (corporeal and incorporeal) and Particle, consciousness would be impossible.

Just as the Theory suggests, Hawkins also believes that memories are stored sequentially (In a "sequence of patterns," he says.) and can only be recalled in the same sequence; however he believes they are all stored in the cerebral cortex ("about 2 millimeters thick and, stretched flat, roughly the size of a large dinner napkin," he writes, and he finds this amazing).

It is the infinite speed of the Particle that permits the amazing ability in man to recall events almost instantly from his past, in great detail (limited only by the relatively slow speed of the physical brain). It is,

Section 11 – Memory and Consciousness

therefore, the Particle that makes possible the thought process and the instant comparisons (not "predictions" as Hawkins prefers) necessary for simple activities like extemporaneous speech.

When someone complains they can no longer remember an event, the memory is still in the Particle. The problem is that their brain, for some reason, cannot recall the event from the Particle. The neural network that placed the memory in the Particle may have been degraded which often occurs with age.

Hawkins knows the brain is too slow to recall memories as fast as we can. He teaches the following: "There is a largely ignored problem with (the) brain-as-computer analogy. Neurons are quite slow compared to the transistors in a computer. A neuron 'collects' inputs from its synapses, and 'combines' these inputs together to 'decide' [53] when to output a spike to other neurons. A typical neuron can do this and reset itself in about five milliseconds, or around two-hundred times per second. This may seem fast, but a modern silicon-based computer can do one billion operations in a second. This means a basic computer operation is five million times faster than the basic operations in your brain! That is a very, very big difference. So how is it possible that a brain could be faster and more powerful than our fastest digital computers? 'No problem,' say the brain-as-computer people. 'The brain is a parallel computer. It

---

[53] Scientists make absurd statements like "A neuron 'collects' inputs from its synapses, and 'combines' these inputs together to 'decide' when to output a spike to other neurons" when they don't know the real answer.

Section 11 – Memory and Consciousness

has billions of cells all computing at the same time. This parallelism vastly multiplies the processing power of the biological brain.'"

Hawkins continues, "I always felt this argument was a fallacy, and a simple thought experiment shows why. It is called the 'one hundred-step rule.' A human can perform multiple tasks in much less than a second. For example, I could show you a photograph and ask you to determine if there is a cat in the image. Your job would be to push a button if there is a cat, but not if you see a bear or a warthog or a turnip. This task is difficult or impossible for a computer today, yet a human can do it reliably in half a second or less. But neurons are slow, so in that half second, the information entering your brain can only traverse a chain one hundred neurons long. That is, the brain 'computes' solutions to problems like this in one hundred steps or fewer, regardless of how many total neurons might be involved. From the time light enters your eye to the time you press the button, a chain no longer than one hundred neurons could be involved."

Hawkins concludes, "So, how can a brain perform difficult tasks in one hundred steps that the largest parallel computer imaginable can't solve in a million or a billion steps? The answer is, the brain doesn't 'compute' the answers to problems; it retrieves the answers from memory. In essence, the answers were stored in memory a long time ago. It only takes a few steps to retrieve something from memory." [54]

---

[54] Think about Jeff's logic here. It does not compute.

Section 11 – Memory and Consciousness

Based on the Theory, it is not the purpose of the brain itself to remember anything. The brain is the principal component of the control and recall process. One of its highest-level roles is to facilitate "focus," allowing the being to concentrate on the most important percepts returned by the Particle. "Memory" is not stored in the brain by any neuron, neural circuit or neural network.

Hawkins, on the other hand teaches, "A typical pyramidal cell (a pyramid-shaped neuron of the neocortex) has several thousand synapses . . . then the neocortex would have roughly thirty trillion synapses altogether. That is an astronomically large number, well beyond our intuitive grasp. It is *apparently* sufficient to store all the things you can learn in a lifetime."

For a time, it was believed memories were stored at the synapses of the brain:

> "Long-term memory is not stored at the synapse," Glanzman [55] said in a press release. "That's a radical idea, but that's where the evidence leads. The nervous system appears to be able to regenerate lost synaptic connections. If you can restore the synaptic connections, the memory will come back."

In other words, the new synaptic connection cannot possibly store the previous memory.

The Theory suggests that not even skills are stored in the physical brain. Nor are spinal reflexes the result of

---

[55] David Glanzman, a neurobiologist at UCLA

Section 11 – Memory and Consciousness

direct "wiring" between sensor and motor neurons. The Particle is involved in all memory-related activities. It takes six hours for the brain to "encode" a new skill, we are told. What is actually taking place is the construction of a neural skill circuit to by-pass the "focus function," so that the being can concentrate on other activities or on the finer points of the skilled activity in process. Spinal reflexes are produced by pre-wired focus-control bypasses, but the neurons involved in any reflex action also receive related, synchronous information from the Particle.[56]

John Ball, editor of *Thinking Solutions* says, "The brain must select the appropriate stored memory to trigger the corresponding rapid muscle movements. It cannot do this with a calculation, as the brain is too slow." That statement supports Hawkins' theory. I will add that, if left up to the brain alone, a dancer could not dance, a tennis player could not play, and you could not carry on extemporaneous discussions. The brain is just too slow and scientists know it.

The Particle has a critical role to play in our lives and it is truly part of you, and will always be a part of you, but it is not you, nor are we "nothing but a pack of neurons," as suggested by Dr. Crick.

No, the brain is not designed to remember anything. I am not the only one who is saying that. Pediatric neurosurgeon, Dr. Michael Egnor says:

---

[56] Note: when the spinal cord above the sensor/motor neuron set is damaged, the brain cannot "modulate" the resulting action and gross reflex actions sometimes occur.

Section 11 – Memory and Consciousness

"A singular consequence of the materialist metaphysics that permeates our culture and our sciences is that we commonly hold basic beliefs that are abject nonsense. One such belief found among ordinary folks as well as neuroscientists, is the belief that the brain "stores" memories. The fact is that the brain doesn't store memories and *can't* store memories. To assert that memories are stored in the brain is gibberish. And don't fall for the materialist invocation of promissory materialism — 'It's just a limitation of our current scientific knowledge, and we promise that science will solve the problem in due time.' The assertion that the brain stores memories is logical nonsense that doesn't even rise to the level of empirical testability."

Here's an example of abject scientific nonsense. Dennis Bray, a Molecular Biologist believes the components of every cell form a micro-computer within each cell which has been programmed by evolution. He explains all of this in his book, *Wetware*. To his credit, Bray recognizes that another, much larger source of information is required, like the Particle, but his answer is ridiculous.

Talk about evolution. Do you remember what I said about Carl Sagan on Page 6? He was a scientist/astronomer and an agnostic. He wrote the book Cosmos and said he couldn't find God anywhere. I wrote to him and said, "No scientist knows how the human cell could possibly contain the enormous amount of information required for human life." He wrote back.

71

Section 11 – Memory and Consciousness

"That is precisely the amount of information gathered over billions of years of chaos. Of course, that only works if humankind has been around for billions of years." It is the atemporal Particle (the human soul) which contains the enormous amount of information required for human life.

[See what Dr. Michael J. Behe, Associate Professor of Biochemistry at Lehigh University, Pennsylvania, has to say about evolution in his fascinating book which deals with the question of the origin of complex biochemical systems, *Darwin's Black Box* [xiii]]

Scientists know there must be a repository of memory, capable of:

1. storing enormous amounts of data, and
2. responding instantly to queries made by the brain.

Francis Crick, in his book called *The Astonishing Hypothesis*, says ". . . memory in the brain has to be stored in a different way" then that of a computer, because the brain is so slow, but he does not share how.

Neurosurgeon, Michael Egnor[57] claims that "It is impossible for the brain to store memories." Yes, he knows that neural damage can cause loss of memory, that certain delicate areas of the brain, if harmed, can destroy the ability to make new memories, and he waves those awkward facts away to announce that "there is

---

[57] Michael Egnor, MD, Evolution News and Views, December 16, 2014

Section 11 – Memory and Consciousness

simply no way memory or information of any kind can be stored in a meat-organ like a brain."

According to the Hypothesis, memories are returned to the same neural network which placed them there in the first place and, therefore, if that neural network is damaged or missing, the memory may not be recallable until and unless the damage is repaired.

Study finds memories stored outside the brain

"The work, published online in the *Journal of Experimental Biology*, can help unlock the secrets of how memories can be encoded in living tissues," noted Michael Levin, Ph.D., Vannevar Bush professor of biology at Tufts and senior author on the paper.

> "As bioengineering and biomedicine advance, there's a great need to better understand the dynamics of memory and the brain-body interface. For example, what will happen to stored memory if we replace big portions of aging brains with the progeny of fresh stem cells?" said Levin, who directs the Center for Regenerative and Developmental Biology in Tufts' School of Arts and Sciences.
>
> Planaria (flat worms) have a remarkable capacity to quickly re-grow new body parts, and decades-old research on planarian learning had suggested that memory can survive brain regeneration. Difficulties inherent in complex and tedious manual worm training experiments

Section 11 – Memory and Consciousness

contributed to planaria falling out of favor as a model for such research, but the new automated training system developed by the Tufts researchers may reverse that.

"We now have a reliable, state-of-the-art approach that moves beyond past controversies to identify quantitative, objective, high-throughput protocols for studying planarian long-term memory capabilities," said Tal Shomrat, Ph.D., first author on the paper. A former postdoctoral associate with Levin, Shomrat is now a postdoctoral researcher at the Hebrew University of Jerusalem. "I believe that investigating this unique animal that displays relatively complex behavior and can regenerate its entire brain in only a few days will provide answers to the enigma of acquisition, storage and retrieval of memories," he added.

Am I wrong? I think they just proved that memories are not stored in the brain.

A brilliant pastor, Fr. Patrick McGarity, told me to keep in mind that, "Along with **error** often comes a small amount of truth." That quote occurred to me as I read the following:

Dr. William Carpenter who, in comparing math prodigy Zerah Colburn's calculating powers to Mozart's mastery of musical composition, wrote the following:

Section 11 – Memory and Consciousness

"In each of the foregoing cases, then, we have a peculiar example of the possession of an extraordinary ~~congenital~~ aptitude for certain mental activity, which showed itself at so early a period as to exclude the notion that it could have been acquired by the experience of the individual. To such ~~congenital~~ gifts we give the name of **intuitions**: ~~it can scarcely be questioned that like the instincts of the lower animals, they are the expressions of constitutional tendencies embodied in the organism of the individuals who manifest them.~~"

I crossed out the error. Congenital means, "existing at birth," while Intuition means, "knowledge apparently acquired without an experience." The key word is "acquired." Intuitive information does not exist at birth, "embodied in the organism of the individual." It must be acquired when the individual is ready for it. I propose that these intuitive "gifts" come from 1) the ancestral memories stored in the Particle and/or 2) are placed in the Particle by the incorporeal brain functioning in the Sequential State. It has already been proven that memory is not stored in DNA nor in the synapse nor in neurons. Ancestral memories are stored in the Particle and some are then experienced as intuitions. How do we get them? Covered later.

Let me conclude with my memory thought experiment:

Do you remember playing on a freshly cut lawn? Did you ever run your hand over the top of the grass? Remember the feeling you got on the palm of your

Section 11 – Memory and Consciousness

hand? Extend out your arm right now with the palm of your hand facing down. Move it back and forth and think about how it felt when you touched the grass. Can you feel the blades of grass touching your palm? Most can.

Consciousness

"So here in lies a choice," says Michael Suede. "You can choose to believe that consciousness is the product of biochemical processes (which is illogical) and that you have no free will, or you can choose to believe that consciousness is eternal, and external to the brain, which allows for free will. No matter what, you cannot say that consciousness is internal to the brain and that you have free will. This is not a logical option." Michael Suede is an Austrian economist and author who holds a business degree from the University of Wisconsin.

"Consciousness at its simplest refers to 'sentience or awareness of internal or external existence'. Despite centuries of analyses, definitions, explanations and debates by philosophers and scientists, consciousness remains puzzling and controversial, being 'at once the most familiar and most mysterious aspect of our lives'. Perhaps the only widely agreed notion about the topic is the intuition that it exists." Wikipedia

So, restated from the Hypothesis unpack, but with added graphics: The Particle is at the functional center of every one of the billions of neurons throughout the Unity. The Particle is communicating continuously with every cell. With the cells in the central nervous system (neurons), there are two-way communications.

Section 11 – Memory and Consciousness

Sensory neurons, such as retinal neurons, transfer information at random to the brain. Cortical neurons in turn communicate this incomprehensible information constantly with the Particle. On a computer screen it would appear as noise.

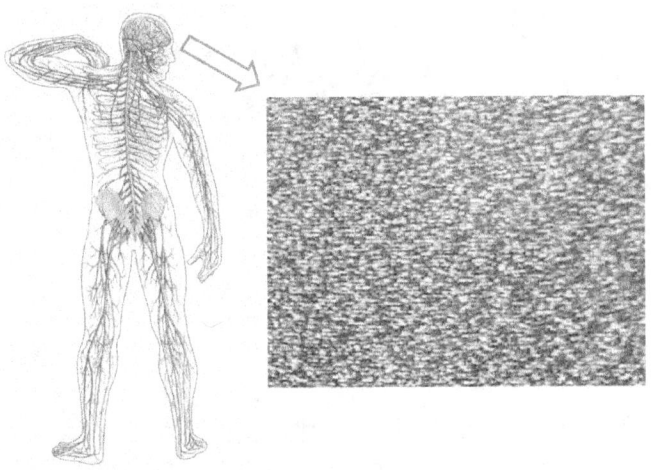

The Particle stores and instantly returns the information received as pulsed, comprehensible percepts, making consciousness possible.

[See next graphic.]

Section 11 – Memory and Consciousness

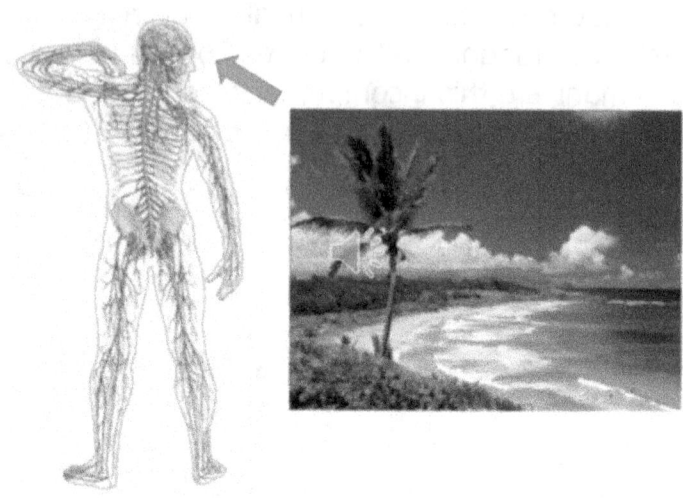

This process is not unlike a television picture refresh process, but the rate is an order of magnitude faster.[xiv]

When the information flow between the Particle and the cortical neurons is caused to stop by anesthetizing the human, they become unconscious to external stimuli, but may still dream.

Consciousness of Vegetative-State Patients

From an article in Medscape:

"Electroencephalography (EEG) may aid physicians in identifying levels of consciousness in patients with limited or no behavioral responses, new research suggests.

For some patients with distinct EEG patterns, fMRI revealed activity that mirrored that of healthy

Section 11 – Memory and Consciousness

volunteers. The activity differed from that of patients in a vegetative or minimally conscious state."

In other words, some patients who did not reveal fMRI activity like healthy volunteers (no blood brain activity), still provided distinct EEG patterns. The Atemporal Particle Theory proposes that the EEG patterns were not produced by the physical brain but were produced by the Particle's Field Signals.

Ultra-High-Speed Recall and Recognition

If the corporeal brain is too slow to recall memories as fast as we can, and as Jeff Hawkins claims, in a "half second (500,000 μs) . . . information entering your brain can only traverse a chain one hundred neurons long," then the memory of the checkerboard image in the following article cannot be recalled and identified by the corporeal brain in 250 μs.

In a study , using high density 256-channel EEG coupled with a liquid crystal display (LCD) tachistoscope to test the brain response to visual checkerboard stimuli for 250 μs (subliminal). Results revealed that checkerboards presented subliminally for 250 μs evoked weak but detectable responses. These results show the presence of a brain response to sub-millisecond subliminal visual stimuli.

The weak but detectable responses came not from the corporeal brain, which is far too slow, but from the incorporeal brain's instantaneous recall (and

Section 11 – Memory and Consciousness

identification) of the checkerboard image from the Particle.

Consciousness and Quantum Mechanics

I am not qualified to discuss how Quantum Mechanics fits into all of this, but I have read enough of the related theories to say there is little doubt in my mind that a physicist may soon make the connection between the Atemporal Particle Theory and Quantum Mechanics. [ii]

I read an article recently on a website called "Ask a Mathematician / Ask a Physicist." The two-part question asked was, "Why are the laws of quantum mechanics so strange? Does it mean that we're missing something?" The questions and a well written response were posted on November 30, 2019 by a contributor called "The Physicist." He wrote, "We're definitely missing something, but we're always missing something. One of the most famous quotes about quantum physics, often used in lieu of a shrug, is due to St. Feynman: "If you think you understand quantum mechanics, you don't understand quantum mechanics."

I responded with (and you won't be surprised): "To answer the primary questions you were addressing, i.e., "Why are the laws of quantum mechanics so strange? Does it mean that we're missing something?" I can tell you what's missing with three words, The Atemporal Particle. I've studied the question, and now the answer, for over 30 years. Scientists will not be able to fully understand quantum mechanics, nor vision, autism,

Section 11 – Memory and Consciousness

consciousness, savant abilities, cellular control and certainly not cancer, until scientists are willing to look beyond the corporeal aspect of life. We already think in terms of massless particles, why not massless, atemporal particles in control of all cellular functions, storing all memory and cooperating in the very ununderstood phenomenon of consciousness."

Section 12
Action Potentials

I thought it would be best, at this point, to provide background on Action Potentials.

The following reference is an excerpt from an article in the September 1992 issue of *Scientific American* entitled, "How Neurons Communicate," by Gerald D. Fischbach.

>   A neuron that has been excited conveys information to other neurons by generating impulses known as 'action potentials.' These signals propagate like waves down the length of the cell's single axon, and are converted to chemical signals at synapses, the contact points between neurons (i.e., contact points on dendrites leading to other neurons).
>   When a neuron is at rest, its membrane maintains an electrical potential of about -70 millivolts (the inner surface is negative relative to the outer

Section 12 – Action Potentials

surface). At rest, the membrane is more permeable to potassium ions than to sodium ions. When the (neuron) is stimulated, the permeability to sodium increases, leading to an inrush of positive charges. This inrush triggers an impulse - a momentary reversal of the membrane potential (known as an action potential). The impulse is initiated at the junction of the cell body and axon, and is conducted away from the cell body.

When the impulse reaches the axon terminals (synapses) it induces the release of neurotransmitter molecules. Transmitters diffuse across a narrow cleft (in the synapse) and bind to receptors in the postsynaptic membrane. Such binding leads to the opening of ion channels and often, in turn, to the generation of action potentials in the postsynaptic neuron.

> **"(Action potentials) have also been traced by fine-tipped micro-electrodes positioned close enough to a (neuron) or an axon to detect the small currents generated as an action potential passes by." GDF**

Hawkins teaches, "You hear sound, see light and feel pressure, but inside your brain there isn't any fundamental difference between these types of

Section 12 – Action Potentials

information. An action potential is an action potential. These momentary spikes are identical regardless of what originally caused them."

The following is proposed by The Atemporal Particle Theory: In all neurons the physical structure of solenoid chromatin fibers, containing DNA, plays a key role in addition to the storing of our coded genetic traits. In all chromatin fibers 1.65 turns of the double helix DNA are wrapped around bead-like octameric histone cores forming nucleosomes. DNA is negatively charged, and the histones are positively charged. The nucleosome assemblies are linked with a short length of the DNA strand. The complete assembly (called chromatin) is twisted tightly forming the solenoid chromatin fiber.[58] Preceding mitosis, it is further coiled, folded and packed into a chromosome.

At rest, the neuron's inner membrane wall is about 70 millivolts negative, relative to the outer wall. When the neuron receives information to be transferred to the Particle, an action potential occurs, reversing the polarization of the cell's membrane walls.

When a neuron of the brain fires, such as after receiving an action potential from another neuron, the electromagnetic pulse is coupled to its nucleus and causes a current in the solenoid chromatin fiber. The electrical characteristics of the solenoid chromatin fiber, produces an electromagnetic field, which propagates the signal to the Particle where it is stored and instantly

---

[58] "Solenoid chromatin fiber" is the scientific term, not just my term.

Section 12 – Action Potentials

returned in the fashion of that described above.

> **The Solenoid Chromatin Fiber, shown below in Figure 4, appears to be an extremely high frequency electronic (electromagnetic) device.**

Signals from the Particle alternately cause an electromagnetic field (Field Signal) which is propagated to the nucleus of the cell. There they may be translated into cellular instructions or, in the neuron, cause the neuron to fire, resulting in the transfer of the information to other neurons as described in Gerald Fischbach's article.

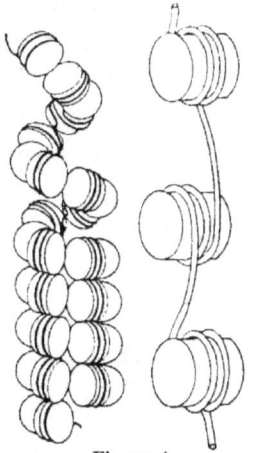

Figure 4.

The Field Signal, originating from the functional center of the nucleus of a cell, envelops the cell and overlaps nearby cells. In non-neural cells, it does not matter if the Field Signal enveloping each cell overlaps adjacent cells, since the communications are generally one way. Regarding neurons, however, an overlap of sufficient signal strength could be problematic, if the Field Signal enveloping one neuron is sufficient to trigger an action potential in an adjacent neuron which may not be in the proper signal path. This may be one reason why neurons are separated as they are by axons leading to synapses connected to dendrites leading to other neurons, etc.

What if, however, in the developing brain of a young

Section 12 – Action Potentials

child, the Field Signal is strong enough, or the neurons are close enough, that the Field Signal enveloping one neuron, under certain conditions, causes an adjacent neuron to fire, which causes another adjacent neuron to fire and so on. A signal "flashover" could occur in a part of the brain, stopped only by a break in the chain-reaction by a lesser Field Signal or greater separation between neurons. This condition may be the cause of "benign childhood epilepsy." It is termed "childhood" because the child "grows out of it." The Theory suggests both the cause for the condition and the reason it does not last into adulthood, both of which are presently unknown.

Data communicated to the Particle is stored as Sequential-State data (defining "Parallata," pl. of Parallatum), i.e., stored with no time in between.[59]

Hawkins also teaches, "The lowest of the functional regions, the primary sensory areas, are where sensory information first arrives in the cortex. These regions process the information at its rawest, most basic level. For example, visual information enters the cortex through the primary visual area, called V1 for short. Your cortex has a primary auditory area called A1 and a primary somatosensory region call S1. Eventually, sensory information passes into 'association areas,' which is the name sometimes used for the regions of the cortex that receive inputs from more than one sense. Most of these areas receive highly processed input from several senses, and their functions remain unclear. The

---

[59] Statements in italics from here out are from the original Truly Astonishing Hypothesis.

Section 12 – Action Potentials

process is generally treated as though information flows in a single direction, but information in the cortex always flows in the opposite direction as well."

The information-storage capacity of the Particle is infinite.

It is the Particle which stores all memory and, as suggested by the Theory, contains all of the information to grow, heal and control every cell – not DNA. The Particle should be considered just a functional part of the human.

Parallata transferred to the Particle are instantly returned (reflected?) to the nuclei of the source neurons. This process converts random data into comprehensible data.

This hypothetical statement appears to address another question that puzzled Dr. Crick. All sensory analog data arriving at different parts of the brain, causing neurons to fire at various rates, left Crick with a few unsatisfactory answers as to how all of this activity is tied into coherent, conscious thought. On the other hand, Hawkins never actually explains how coherent, conscious thought results from the two-way flow of information he believes is always occurring.

The eyes, in truth, are "the windows of the soul," because the Particle is located at the functional center of all retinal neurons.

The sensory neurons of the retina receive analog

Section 12 – Action Potentials

information coming through the lens of the eye.

Neurons of the brain receive and transfer this analog information to the Particle, and it is returned to the brain in a synchronous, comprehensible manner. It is, therefore, the communication of information between the brain and Particle which makes vision possible.

The eyes, in truth, are the "windows of the soul." The Particle, in fact, is involved in every movement we make, every sight we see, every smell, every sound and every touch. We are living beings only if the cortical neurons of the brain, corporeal and incorporeal, and the Particle are communicating. Together, the source of information stored by the Particle and the neurological system of the brain form the "mind," an intellectual tool controlled and focused, when we are conscious, by the "body-soul-spirit" [60] triunity, which we are.

Brain cells communicate with the Particle every time a neuron generates an action potential by collapsing and rebuilding a voltage field surrounding the DNA within the nucleus of the cells. In turn, the DNA, coiled as it is into a solenoid configuration, produces a very high frequency electromagnetic pulse which envelops the Particle.

---

[60] "spirit" is defined later.

Section 13
Reaction to unexpected events

As Jeff Hawkins says, the brain is too slow to compute reactions to stimuli. As an answer to this problem he says the brain retrieves the answers from memory (ergo, the Particle. He's right). Other scientists say the brain can react to stimuli in about 300 milliseconds (a little more than $1/4^{th}$ of a second). You can test this online, if you like. For example, you will be asked to touch a key when you see a red circle, versus a blue triangle, etc. But you will be watching for a red circle, an expected event. Your reaction to an unexpected event is even faster.

You may be working at your desk, focused on what you must do, or reading an email, or watching a video when quite suddenly a pencil falls off the corner of your desk due to an unrelated motion of yours. You were not watching the pencil; your focus was on your work. Yet, somehow, in less time than it takes to turn your attention toward the falling pencil, your hand reaches for it, grasps it, catching it in midair within inches of your

Section 13 – Reaction to unexpected events

desktop without you even thinking about it. How is that possible, if the brain is so slow?

Here's what other scientists think has happened. You saw the pencil in your far peripheral vision where the density of receptor cells in the retina is lowest at the edges. This information traveled through an electro-chemical process along axons (your first hint of low speed – not the speed of electrons moving through a conductor[xv]) to the synapses (a chemical exchange of neurotransmitters – your second hint of slow speed) leading to the dendrites of visual cortex neurons which take 5 milliseconds (more slow speed) to create an action potential (a voltage swing across each neuron's outer membrane of .10 millivolts). The action potentials travel on axons to the synapses (another time-consuming chemical transfer of the information) connecting to the dendrites of your cortical neurons. Somehow, your collection of cortical neurons (requiring at least 5 milliseconds each to act) know what is happening from a memory in the brain (according to Jeff Hawkins – we'll ignore how much time that requires) and send action potentials down their axons to the synapses leading to the dendrites of a multitude of motor neurons. Finally receiving the information, the motor neurons fire (in another 5 milliseconds) and cause the exact muscles to contract, shaping the fingers of the hand correctly to grasp, in an instant, the falling pencil.

Here's what the Atemporal Particle Theory suggests really happens.

Section 13 – Reaction to unexpected events

You saw the pencil in your far peripheral vision where the density of receptor cells in the retina is lowest at the edges. The Particle at the functional center of these neuron-receptors instantly (no time involved) compares the information to a similar reaction in memory and instantly causes action potentials in the appropriate motor neurons (a delay of 5 milliseconds) which in turn cause the exact muscles to contract, shaping the fingers of the hand correctly to grasp, in an instant, the falling pencil.

And how, you may ask, does the Particle communicate directly with the motor neurons? Just like it does with all neurons in the body. The Particle and DNA are at the functional center of every cell in the body, with the exception of red blood cells. All cells receive instructions from the Particle, but neurons and the Particle share information. How does the transfer of information from the neuron to the Particle take place? DNA is wrapped 1.65 turns around bead-like octameric histone cores forming nucleosomes [61]. The assembly is twisted tightly forming solenoid chromatin fibers.

At rest, the neuron's inner membrane wall is about 70 millivolts negative, relative to the outer wall. When the neuron receives information to be transferred to the Particle, an Action Potential occurs, reversing the polarization of the cell's membrane to +40 millivolts.

The Action Potential thereby creates a current in the solenoid chromatin fiber within the nucleus of the

---

[61] Tell that to an electronics engineer and watch their reaction.

Section 13 – Reaction to unexpected events

neuron which produces a very high frequency electromagnetic field enveloping the Particle.

The reverse happens when the Particle shares information with the neurons, in this example, motor neurons. The Particle at the functional center of the motor neuron produces a very high frequency electromagnetic field enveloping the solenoid chromatin fibers which act like a fractal antenna.[62] This produces an action potential which leaves by way of the neuron's axons traveling to the muscles causing contractions.

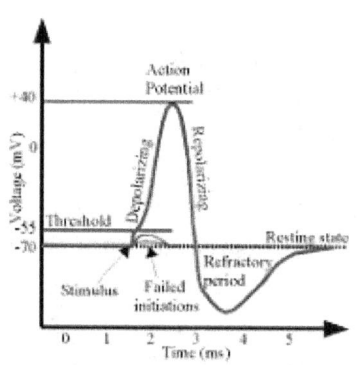

As I've said before, if it were not for this process, you could not dance, play tennis, baseball or even carry on an extemporaneous conversation.

[See the graphic on the next page.]

---

[62] "The wide frequency range of interaction with EMF is the functional characteristic of a fractal antenna, and DNA appears to possess the two structural characteristics of fractal antennas, electronic conduction and self-symmetry." Martin Blank and Reba Goodman, Journal of Radiation Biology, Volume 87, 2011– Issue 4

Section 13 – Reaction to unexpected events

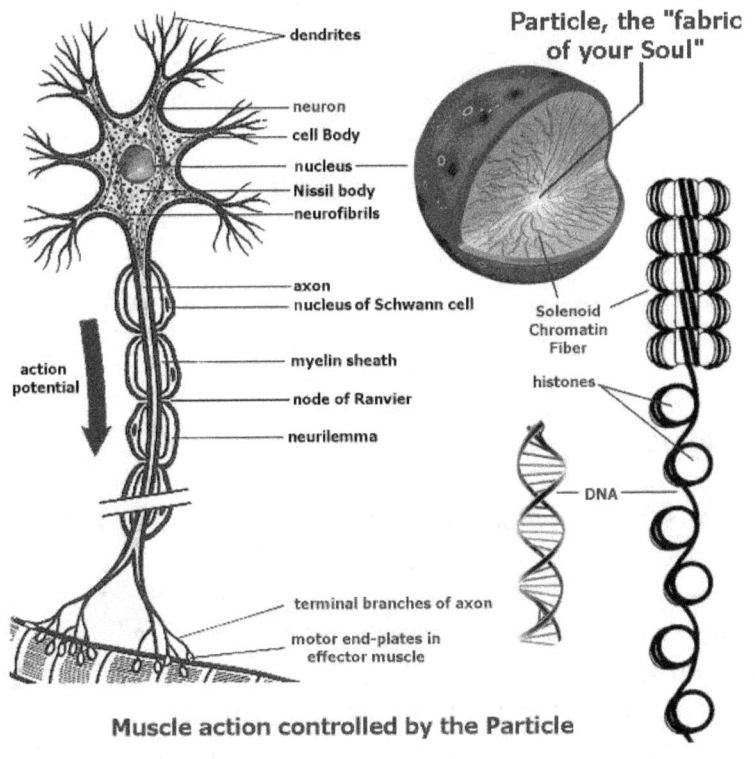

Section 14
Reproduction

Mitosis

Mitosis (cell division) is the process during which a single cell divides into two identical daughter cells. "It has long been known that mature, differentiated neurons do not divide" (National Center for Biotechnology Information), therefore, the process of Particle/neuron communications is not interrupted by mitosis for neurons. However, Particle instructions to cells that do divide would be interrupted during DNA replication if it were not for the centrosomes. Centrosomes consist of two centrioles positioned at 90 degrees to one another.[63] They also have high-frequency electrical characteristics (see the graphic on the following page) and therewith receive instructions from the Particle.

Centrosomes are located just outside of the nucleus of the cell. During the Prophase of the cycle, the nucleus

---

[63] Two antenna-like devices placed at 90 degrees to one another? Tell that to an electronics engineer and see what they say. Unfortunately, only biologists know about centrosomes.

Section 14 – Reproduction

disappears, and the centrosomes are positioned on both sides of the cell.

Conception

At the true moment of conception, the Particles of both mother and father unite creating a new, unique Particle incorporating the instructions to develop a unique Unity. The Particle, therefore, also contains the combined ancestral memories of the parents. Concurrently at conception, the corporeal chromosomes and the incorporeal chromosomes of both parents unite within the ovum creating "a genetically unique, newly existing,

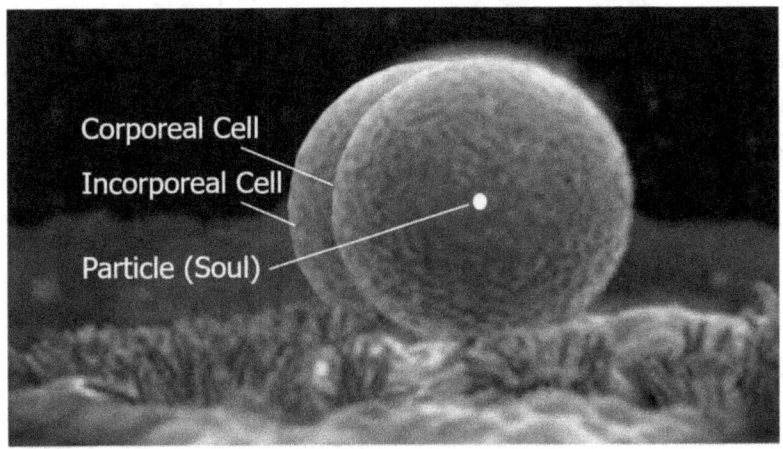

Section 14 – Reproduction

individual, a whole living human being as a (dual) single-cell embryonic human zygote." [64]

Corporeal chromosomes, through which the Particle communicates, carry imperfections and errors which may have developed ancestrally. Incorporeal body Chromosomes, which communicate simultaneously with the same Particle, do not carry imperfections and errors. External radiation can interfere with the communications between the Particle and corporeal body cells; and therefore, the information being transferred to/from the cells may be affected, causing errors in communications especially significant during cell division. [Placing a cellular phone close to your head or body may not be a good idea.]

At the obvious risk of criticism from Christians, but with all due respect, I will show how the God/Man Jesus was conceived as fully God while at the same time being fully human, applying the Atemporal Particle Theory.

Jesus' Incarnation

As I said above, all humans begin as a single corporeal cell, coincident with an incorporeal cell, both with one combined Particle from their parents. The Particle controls the growth and function of all cells, corporeal

---

[64] (International Journal of Sociology and Social Policy 1999, 19:3/4:22-36 (in press) WHEN DO HUMAN BEINGS BEGIN? "SCIENTIFIC MYTHS AND SCIENTIFIC FACTS" Dianne N. Irving, M.A., Ph.D. (copyright February 1999)

Section 14 – Reproduction

and incorporeal, and it is the repository of all memory and the co-generator of consciousness. Jesus' incarnation was no different, with certain major exceptions.

Somehow, God may have caused Mary's ovary to produce, through an anomaly during meiosis, a gamete (ovum) with 46 chromosomes, not just 23, which would include an X and a Y chromosome. Such an anomaly is possible, but an extremely rare event in humans. Or, also through an extremely rare event in humans, Mary may have produced a gamete with an X chromosome and another gamete with a Y chromosome which then fused into a male bipartite (corporeal/incorporeal) zygote.

The Holy Spirit then caused the incorporeal cell of Jesus God to replace, or exist in place of, the incorporeal cell of Mary's zygote and become coincident with the corporeal cell produced by Mary. Keep in mind that God is not limited by time nor the sequence of events. Thereby, Jesus, a fully human and fully God male, began to develop, from a single bipartite cell in the womb of Mary with a Particle coming only from His mother. The coincident cells, therefore, developed into the person of Jesus with a corporeal human Body and a coincident divine incorporeal body.

Upon His birth, as with all humans, Jesus' corporeal body's brain became dominant, and because His Particle came only from Mary, He was unable to recall His divine nature. I assume this was Jesus' intention. "He emptied

Section 14 – Reproduction

himself, taking the form of a slave, coming in human likeness; and found human in appearance." [65]

The chromosomes of Jesus' corporeal body, through which his Particle communicated, carried imperfections and errors which may have developed ancestrally. His corporeal body, therefore, was not a perfect likeness of the replaced incorporeal cell from his mother, which had no imperfections. "He grew up like a sapling before him, like a shoot from the parched earth; He had no majestic bearing to catch our eye, no beauty to draw us to him." [66]

[I've published a booklet called, "Jesus – A Scientist's Perspective" which goes much further into detail. It is available through most online bookstores.]

---

[65] Philippians 2:7
[66] Isaiah 53:2

Section 15
Life and Healing

Here are some excerpts that will serve as good background:

"Stem Cells: A Primer"

Human development begins when a sperm fertilizes an egg and creates a single cell that has the potential to form an entire organism. This fertilized egg is **totipotent**, meaning that its potential is total. In the first hours after fertilization, this cell divides into identical totipotent cells. This means that either one of these cells, if placed into a woman's uterus, has the potential to develop into a fetus. In fact, identical twins develop when two totipotent cells separate and develop into two individual, genetically identical human beings.[67] Approximately four days after fertilization and after several cycles of cell

---

[67] See Twins on the following pages.

Section 15 – Life and Healing

division, these totipotent cells begin to specialize, forming a hollow sphere of cells, called a blastocyst. The blastocyst has an outer layer of cells and inside the hollow sphere, there is a cluster of cells called the inner cell mass.

The outer layer of cells will go on to form the placenta and other supporting tissues needed for fetal development in the uterus. The inner cell mass cells will go on to form virtually all of the tissues of the human body. Although the inner cell mass cells can form virtually every type of cell found in the human body, they cannot form an organism because they are unable to give rise to the placenta and supporting tissues necessary for development in the human uterus. The inner cell mass cells are **pluripotent** - they can give rise to many types of cells but not all types of cells necessary for fetal development. Because their potential is not total, they are not totipotent, and they are not embryos. In fact, if an inner cell mass cell were placed into a woman's uterus, it would not develop into a fetus.

The pluripotent stem cells undergo further specialization into stem cells that are committed to give rise to cells that have a particular function. Examples of this include blood stem cells which give rise to

Section 15 – Life and Healing

red blood cells, white blood cells and platelets; and skin stem cells that give rise to the various types of skin cells. These more specialized stem cells are called **multipotent**.

A primary goal of (stem cell research) would be the identification of the factors involved in the cellular decision-making process that results in cell specialization. We know that turning genes on and off is central to this process, but we do not know much about these "decision-making" genes or what turns them on or off. [68] Some of our most serious medical conditions, such as cancer and birth defects, are due to abnormal cell specialization and cell division. A better understanding of normal cell processes will allow us to further delineate the fundamental errors that cause these often-deadly illnesses. Source: National Institutes of Health - May 2000

If people knew that at conception the Particles (souls) of the partners are also combined, not just DNA, will they look differently on abortion?

Twins

As stated in the Stem Cell Primer, identical twins develop

---

[68] The answer, of course, is that the Particle controls the genes.

Section 15 – Life and Healing

when two totipotent cells separate and develop into two individual, genetically identical human beings. Such twins begin life with the same Particle, the combined Particles of their parents. They are not at all differentiated until they begin to have separate experiences. Remember, they each have Sequential State bodies and brains that are also identical. It is well known that identical twins often finish each other's sentences. I watched a video of two girls, identical twins, who have seldom been apart, having had all of the same experiences. When they talked with an interviewer, their answers to questions often sounded like they were both speaking as one. There was only a barely perceptible echo when one began to say the same thing as the other. I'm going to call this Simultaneous Intuition, and it leads to an explanation of what intuition is and how it occurs.

Intuition

First, an example: If you were standing in front of your well-organized shelf of books (mine is not), and you had the "sure feeling" that somewhere in the 175 books on the shelf was one covering the current subject of your interest, you may have been receiving an intuition. You could not recall such a book, but in your "subconscious mind" you knew if you would search through all of the books, you would find the book of your interest. After a book-by-book search, you find it! From whence did the intuition come?

The instant your Temporal State eyes and brain were turned toward the bookshelf, in hopes of finding the

Section 15 – Life and Healing

book, your Sequential State eyes and brain scanned all 175 of the book titles on the shelf and instantly found the book of interest to you. The knowledge and image of the book found was automatically placed in the Particle by your Sequential State brain, and when your Temporal State brain queried the Particle for a memory of such a book, you got the feeling the book was somewhere on the bookshelf. An intuition.

Why couldn't you see the image of the book binding placed there by your Sequential State brain? Because your Temporal State brain did not store the image in the Particle and therefore it could not be recalled to the exact same neural network which placed it there. On the other hand, you have a similar neural network in your Temporal State brain, so you were at least able to recall the memory storage action of your Sequential State brain.

In short, an intuition is a memory in your Particle of something your Temporal State Brain did not experience. Dr. William Carpenter was right when he wrote, intuition is "knowledge apparently acquired without an experience." [69]

Do not lose sight of the fact you are at once a Temporal State body and Sequential State body. They are both you.

Back to the twins and Simultaneous Intuition. The two twins I described above were seldom apart and

---

[69] See Page 69.

Section 15 – Life and Healing

therefore, had almost identical experiences, placed, therefore, in almost identical Particles (or was it the same Particle?). As one spoke, the other instantaneously received an intuition of what was going to be said by her sibling and she said the same thing through Simultaneous Intuition.

Growth and Healing

Growth and healing[xvi] are the same process and are similarly directed by the Particle. When an organism is injured or diseased, and cells are damaged or destroyed, it is the Particle that provides the correct code (Likeness) with which to direct the physical regeneration process when possible. Note: Incorporeal cells cannot be injured or become diseased.

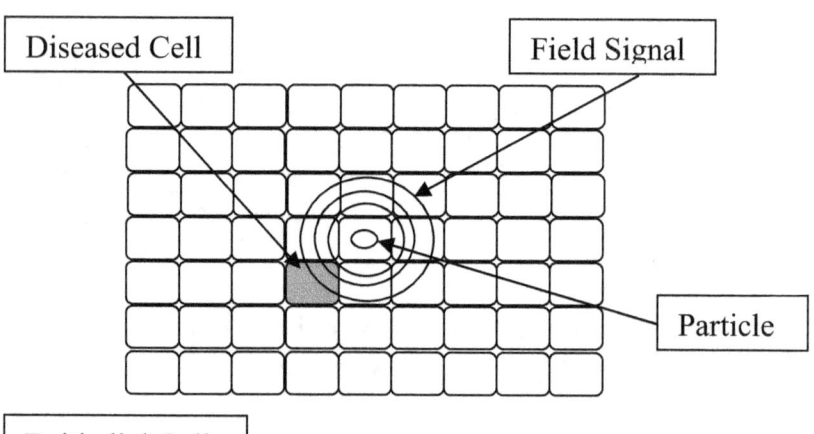

Each living cell, corporeal and incorporeal, receives specific instructions from the Particle, including instructions to divide, when necessary, for the growth or

Section 15 – Life and Healing

healing of the organism. If a cell is missing, diseased or sufficiently damaged, it cannot receive these instructions, nor can it in any way report its condition. Instructions to divide for the growth or healing of the organism must be given to a healthy cell adjacent to or in line with (heretofore defines "Corresponding") the position of a missing, diseased or damaged cell.

The Field Signal, carrying the instructions from the Particle to the cell, radiates from the cell, reaching the nuclei of Corresponding cells. There the cell-specific information is forwarded by the Corresponding-cell nuclei (i.e., by the structure of the chromosomes or Solenoid Chromatin Fiber) to the Particle. Non-reporting cell positions are identified in this manner. When the need for cell division is recognized, and a divide instruction is sent, other Corresponding cells are inhibited from dividing by the cell-specific instructions radiating from the dividing cell. In this way the growth or healing process is controlled.[xvii]

When a cell receives instructions from the Particle, and existing, even healthy Corresponding cells do not, for any reason, forward the cell-specific instructions carried by the Field Signal; the originating cell may be instructed to divide even though there is no need for division. Since a cancer cell does not recognize proper instructions regarding shape, purpose, color, structure, location or division limits, it can be assumed it may also be incapable of being a source for correct cell-specific instructions to Corresponding cells. Thus, no Corresponding cells are recognized by the Particle, and

Section 15 – Life and Healing

the cancer cell continues to receive instructions to divide. [70]

Although the DNA code of a cancer cell remains identical to a healthy cell, the physical structure of some of the cell's DNA may have been altered or damaged in some way.[xviii] Many cancers are associated with a chromosome defect in which part of one chromosome appears to have broken off and joined with another chromosome. This improper assembly of a chromosome could clearly occur during the cell-division process (mitosis – see Page 94). Interference with the Field Signal during mitosis may cause the dividing cell to misinterpret critical instructions from the Particle. [xix]

As explained earlier, during mitosis it appears to be the centrosome that is receiving cell division instructions from the Particle, since the structure of DNA is disassembled during the process. "The centrosome consists of two structures called centrioles set at right angles to each other and are surrounded by a cloud of pericentriolar material. Seen in cross section, a centriole reveals a pinwheel structure made of structural elements called microtubules" (*Scientific American*, June 1993 – David M. Glover).

---

[70] It can also be assumed that cancer cells, corresponding to healthy cells, are incapable of receiving and reflecting the Particle's cell-specific information from the healthy cells, and are thereby identified by the Particle as non-reporting cell locations, causing even the healthy cells to divide. This could multiply the rate of unwarranted growth in the area of the cancer cell.

Section 15 – Life and Healing

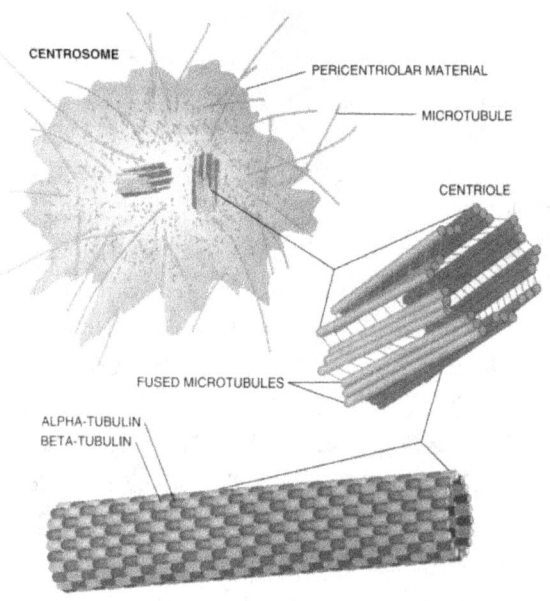

The American Cancer Society says, "Cells become cancer cells because of damage to DNA. DNA is in every cell and directs [71] all its actions. In a normal cell, when DNA gets damaged,\ the cell either repairs the damage or the cell dies. In cancer cells, the damaged DNA is not repaired, but the cell doesn't die like it should. Instead, this cell goes on making new cells that the body does not need. These new cells will all have the same damaged DNA as the first cell does." . . . and, therefore, the new cells will also misinterpret instructions from the Particle.

---

[71] DNA is just a chemical and cannot by itself "direct" cell actions. It is the Particle which has that information.

Section 15 – Life and Healing

"In recent years one of the intriguing discoveries has been that while one part of the DNA polymerase molecule functions as a polymerase, attaching nucleotides, another portion of the same molecule acts as an "exonuclease" (nucleotide cutting-out enzyme) and performs a "proofreading" function. It is estimated that about one time in 10,000 to 100,000 the wrong base is added to a growing DNA strand. *Somehow* the exonuclease portion of the DNA polymerase molecule recognizes nearly all such mistakes and removes each erroneous base as fast as it is added so that another attempt to add the correct one can be made. The result is that there is an estimated error rate of only one in one billion base pairs copied (during mitosis)."

The above paragraph was taken from the *Biology Coloring Book* by Robert D. Griffin. The "*somehow*" is explained by the Theory. Based on the Theory, if the communications between the Particle and the cell were perfect, no errors would result (or remain). Gross mistakes causing the displacement of part of a chromosome could also result in repeating errors in reception of information from the Particle through the Field. Such flaws would also affect the ability of cells to

Section 15 – Life and Healing

"reflect" cell-specific information back to the Particle as described in later paragraphs. [72]

Cell division is normally inhibited if nutrient levels are below certain limits. Cancer cells (and possibly healthy adjacent cells), on the other hand, will divide even when nutrient levels are one tenth of this limit. The reason may be that the Particle recognizes the non-reporting cell positions as an injury to the organism, not just normal cell replacement as in the case of surface skin cells.

Though the signal strength of the Field Signal may drop in direct proportion to the square of the distance from the functional center of the nucleus from which it is originating, the Field Signal reaches more than just the Corresponding cells or even cells of the same type. In this manner the Particle receives an encoded "picture" of the entire organism. So it is with the epithelial cells of the skin. The Particle receives a precise thickness measurement when it sends cell-specific information to a germinating layer cell through the Field Signal, which radiates to (and probably beyond) the unattached surface layer cells.

If a missing, diseased or damaged cell is beyond the physical range of the Field Signal radiating from a healthy cell, it is not recognized as a non-reporting cell location because no cell is expected within the range

---

[72] Do some wavelengths of Electromagnetic Radiation interfere with the Field Signals? Probably.

Section 15 – Life and Healing

covered. Through action potentials, neurons are capable of transferring information from other neurons to the Particle by means of the Field Signal and from the Particle to other neurons. In part, neurons are spaced to avoid causing an action potential in an adjacent neuron when its neighboring neuron generates a data-carrying Field Signal, which also overlaps the adjacent neuron. This requirement makes the strength of the radiating Field Signal more critical with regard to neurons, especially in adults, when it comes to the process of recognizing adjacent, damaged, diseased or missing neurons.

The tree-like structure of cells forming the central nervous system, versus the layered structure of epithelial cells, further complicates cell replacement, and the healing of large gaps in the structure from injury, disease or other causes of cell death would seem to require a different process, guided by the Particle. However, when new neurons are surgically deposited in these gaps the cell replacement process described above would be facilitated.

[Note: Perhaps the body needs sleep because the thought process (consciousness) occupies the temporal/atemporal "link" during waking hours to the extent that non-REM sleep periods are required for communications involving cell functions and healing. Perhaps, also, in between these regeneration communications the "system" releases the mind temporarily, which begins to dream. ^{xx}]

I think you will agree that God does not need to use magic to heal an individual we are praying for. He uses

Section 15 – Life and Healing

the means He provided for in His Creation. If we ask God to heal a child born with a club foot, and He does so almost instantly, based on the Theory, what has actually happened? The Particle, in the Sequential-State, in which events happen sequentially with no time in between, directs the cells of the child's foot to divide and grow according to the perfect Likeness stored in the Particle. The resulting metabolic activity no doubt generates heat, which the child would experience, as the blood vessels carry the heat throughout their body. Knowing this, you can prepare the individual you are praying for, increasing the individual's believe that God is actually going to heal them, and we all know that faith is necessary for such healings. The individual must accept (will) what is about to happen.

Note: For the Atemporal Particle Theory on cancer see End Notes, Page 179.

Section 16
Aging

We were designed to live forever in health and happiness. How could it be any other way?

Leonard Hayflick, Ph.D., a professor of anatomy at the University of California, San Francisco, says, "The accumulation of new insights has made it possible, for the first time, to understand the biological reasons for the aging of animals and humans. Aging occurs because the complex biological molecules of which we are all composed become dysfunctional over time as the energy necessary to keep them structurally sound diminishes. Thus, our molecules must be repaired or replaced frequently by our own extensive repair systems," Hayflick said. "These repair systems, which are also composed of complex molecules," he explained, "eventually suffer the same molecular dysfunction. The time when the balance shifts in favor of the accumulation of dysfunctional molecules is determined by natural selection[73] and leads to the manifestation of

---

[73] Does not refer to evolution.

Section 16 – Aging

age changes that we recognize are characteristic of an old person or animal. It must occur after both reach reproductive maturity, otherwise the species would vanish." Hayflick also noted that these repair and maintenance systems are called "determinants of longevity," which is a phenomenon different from the aging process itself. "These fundamental molecular dysfunctional events lead to an increase in vulnerability to age-associated disease," he said. "Therefore, the study, and even the resolution of age-associated diseases, will tell us little about the fundamental Processes of aging."

Man's Temporal body was designed to live forever. It is the imperfect communications between the Particle and Temporal State cells that gradually bring aging to the organism.

"Aging occurs on the cellular level in a number of ways. The strands of DNA that guide a cell's physiological process (through instructions from the Particle) can be damaged during the normal function of generating RNA, the cell's messengers. The presence of repair mechanisms (the polymerase molecule directed by the Particle) substantiates this notion, because if the DNA was not properly repaired, it would impair cellular functions. [74] Thus, it is thought that by either wear and tear or through improper repair over time the DNA could be destroyed and the cellular reproduction function impaired." (*Grolier Multimedia Encyclopedia*)

---

[74] and instructions from the Particle.

Section 16 – Aging

In an article by astronomer Dr. Hugh Ross, he writes, "Medical experts agree that cosmic radiation plays a significant role in limiting human life spans. Astronomers agree that the vast majority of this life-limiting radiation comes from supernovae, cataclysmic explosions of super giant stars." The article goes on to say that astronomers Erlykin and Wolfendale confirmed that the Vela supernova is indeed the prime contributor.[75]

The Theory confirms much of what is known about wellness and long life. Adequate sleep is necessary. During non-REM sleep communications between the Particle and cells are uninterrupted by conscious thought which may fully occupy the "system" during waking hours. Eating healthy foods maintains healthy cells and provides the chemical components for cell reproduction. Undernourished cells, including neurons, may not be as able to receive, decipher and carry out instructions from the Particle relative to the Likeness. These neurons may not be able to produce Field Signals or respond as quickly (or fully) to the Field Signals produced by the Particle, therefore cognitive powers and memory-recall will be affected.

Avoiding radiation takes on a whole new meaning when the Field is taken into consideration. We know that ultraviolet radiation damages the DNA structure of cells. Instructions from the Particle will be either not received, will be misinterpreted or not acted upon. Dr. Michael Reacholi, manager of the WHO's *Electromagnetic Fields*

---

[75] Which some say occurred about the time life spans were limited to 120 years.

Section 16 – Aging

Project, told a news conference that, "There are key issues that still need to be resolved because there have been suggestions that electromagnetic fields may produce cancers or memory loss or other neuro-degenerative diseases." The Theory suggests, using a cellular phone with its antenna close to the brain is probably not a good thing to do. Living near low-level electromagnetic radiation (EMR) from power lines over a long period of time may not be safe. [76] The Theory also suggests that EMR could disrupt communications from/to the Particle. However, there will be no direct cause/effect found until the existence of the Field Signals are recognized and the Particle "discovered."

Aging is not part of God's plan for us. We brought it on by turning from Him and disobeying Him. All of nature began to fail at that first moment, and communications between our souls (Particles) and cells fail on occasion as a result. Aging can also be accelerated by our own personal actions.

---

[76] Coincidently, my wife, Kim's parents lived near high-tension powerlines for most of their lives and both developed cancers. Kim does not believe it was coincidental.

Section 17
Death - Transition

Neither Crick nor Hawkins say anything about death. In the index of their books, death is not listed. Why not? Death, or "transition" as I prefer to call the process, is a natural part of our existence. There's much interest. A search on Google.com for articles on death returned 341 million results.

We can't talk about transition (physical death) without discussing near-death experiences. Dr. Michael Sabom is a cardiologist in private practice who has studied near-death experiences for over 20 years. He reports as a scientist about what he has learned by interviewing nearly 50 individuals who "returned from death's door."

Dr. Sabom's book, *Light and Death*, shares with the world his findings. Sabom, also now a born-again Christian, scrutinizes near-death experiences in light of what the Bible has to say about death and dying, the realities of light and darkness, and the gospel of Jesus Christ.

Section 17 – Death-Transition

In his book, Dr. Sabom relates an interview with a surgery patient he named "Pam Reynolds," a 35-year old woman, who said she was operated on for a giant artery aneurysm in her brain. She told him she underwent a dangerous surgical procedure nicknamed, "Standstill," because they cooled her body to 60 degrees, stopped her heart and drained all of the blood from her brain. She was by all clinical standards, quite dead. During that time, Pam had an Out-of-Body Experience, which you read about earlier.

When Dr. Michael Sabom heard Pam describe the bone saw that the surgeon used to open her skull, "The 'saw' thing . . . looked like an electric toothbrush," he said "No way," and filed away the interview tape. A year later he decided to research her story. He called the company that made such saws and they sent him a student's user manual with pictures. He wrote, "I was shocked by the accuracy of Pam's description of the bone saw as an electric toothbrush and with the 'socket wrench case,' in which the equipment is kept." However, she said there was a groove at the top of the saw and the initial picture I had showed no such groove. Another image of the saw blade later matched her story.

Pam also said she heard the cardiologist say that her veins were small. Sabom contacted Pam's surgeon who told him it was the cardiovascular surgeon who commented about the small veins in her report, and that Pam could not have heard her speak because Pam's ears were covered with tightly fitting earphones that produced a loud clicking noise and blocked all outside sound. Nonetheless, Pam's description of what was said

Section 17 – Death-Transition

was accurate. There are thousands of such experiences recorded, some personally experienced by doctors, scientists and even an airline captain.

P.H.M. Atwater carefully reviewed many near death experiences and saw a similarity that could not be ignored. Here is what she found:

- Most felt a sensation of floating out of one's body . . . where all that goes on around the vacated body is both seen and heard accurately (like Pam Reynolds reported).
- Passing through a dark tunnel or dark hole or encountering some kind of darkness. This is often accompanied by a sensation of moving or acceleration.
- Ascending toward a light at the end of the darkness; a light of incredible brilliance, with the possibility of seeing people, animals, plants, lush outdoors and even cities within the light.
- Greeted by friendly voices, people or beings who may be strangers, loved ones or religious figures. Conversation can ensue; information or messages may be given.
- Seeing a panoramic view of the life just lived, from birth to death or in reverse order.
- A reluctance to return . . . but invariably realizing or being told their job on earth was not finished or a mission must yet be accomplished before they can return to stay.
- A warped sense of time and space; discovering time does not exist.

Section 17 – Death-Transition

- And some experience a disappointment at being returned.

Now let's take our study to the next step. What happens when we die? Do we die? What does the Christian Bible have to say about death? Keep the words and phrases I have underlined in the following verses in mind as you read the following hypothetical statements.

Jesus said, "I am the resurrection and the life. Anyone who believes in me will live, even <u>though they die</u>; and whoever lives by believing in me <u>will never die</u>." John 11:25 (TNIV)

1 Thessalonians 5:23 (TNIV) "May God himself, the God of peace, sanctify you through and through. May your whole <u>spirit</u>, <u>soul</u> and <u>body</u> be kept blameless at the coming of our Lord Jesus Christ."

John 11:11-14 (TNIV). "After he had said this, he went on to tell them, 'Our friend Lazarus has fallen asleep, but I am going there to wake him up.' His disciples replied, 'Lord, if he sleeps, he will get better.' Jesus was speaking of his <u>death</u>, but his disciples thought he meant <u>natural</u> sleep. So, then he told them plainly, 'Lazarus is dead.'

When they arrived at the tomb, Jesus looked up and said, 'Father, I thank you that you have heard me. I know that you always hear me, but I said this for the benefit of the people standing here, that they may believe that you sent me.' When he had said this, Jesus called out in a loud voice, 'Lazarus, come out.' The <u>dead</u>

Section 17 – Death-Transition

man came out, his hands and feet wrapped with strips of linen, and a cloth around his face." John 11:41-43 (TNIV).

Luke 8:51-55 (TNIV). "When he arrived at the house of Jairus, he did not let anyone go in with him except Peter, John and James, and the child's father and mother. Meanwhile, all the people were wailing and mourning for her. 'Stop wailing,' Jesus said. 'She is not <u>dead but asleep</u>.' They laughed at him, knowing that she was dead. But he took her by the hand and said, 'My child, get up!' Her <u>spirit</u> returned, and at once she stood up. Then Jesus told them to give her something to eat."

Luke 23:46 (TNIV) "Jesus called out with a loud voice, 'Father, into your hands I commit my <u>spirit</u>.' When he had said this, he breathed his last."

Humans are body-soul-spirit beings (triunities) during their temporal life.

Without the soul the human does not exist. At the same time, without Field Signals they would not know they exist, and their physical bodies would die, because their atemporal souls (Particles), in the Sequential State, could not communicate with their bodies in the Temporal State.

At the moment of the medically-defined death of an individual, which occurs when Particle/cortex two-way communications are interrupted (for whatever reason), the atemporal, Sequential-State body leaves the Temporal-State body.

Section 17 – Death-Transition

The Sequential-State body/soul unity is the "Spirit" of the individual. Defining Spirit.

The Spirit leaves the Temporal-State body entering the Out-of-Body Stage of the "dying process."

The next stage of the dying process involves the Spirit being transported out of the physical realm. This is the Near-Death Stage of the dying process.

If the cause for Particle/cortex two-way communication interruption is reversed, the Spirit can still return to the physical realm and from the Near-Death Stage, and the Temporal-State body can again be animated by the Particle (resuscitation).

Surgeon Bernie Siegel gives this account in his book, *Love, Medicine and Miracles*: "Once, as I finished a difficult emergency abdominal operation on a young, very obese man, his heart stopped just as we were about to remove him to the recovery room. He didn't respond to resuscitation. The anesthesiologist had given up and was walking out the door when I spoke out loud into the room, 'Harry, it's not your time. Come on back.' At once the cardiogram began to show electrical activity, and the man ultimately recovered fully. I can't prove it, of course, but I'm sure the verbal message made the difference. I know the experience made believers out of the other staff members who were present."

If the cause for the Particle/cortex two-way communication interruption is not reversed and the

Section 17 – Death-Transition

Corruption [77] of the Temporal-State body begins, the Spirit cannot return and the death process is irreversible. When the dying process is complete, the individual, as a Spirit (Sequential-State body/soul Unity), goes to God's presence – the Eternal Stage. "Father, into thy hands I commend my Spirit." Luke 23:46

When the dying process is complete, the individual, as a Spirit, goes to God's presence. "As a Spirit" does not mean as an ethereal, vapor-like, unrecognizable ghost. The individual will look, to others in the spiritual state, as they did to us in the Temporal-State (except probably younger), and they will "physically" feel great. They will be able to see, hear and probably touch those they meet just as they did in the Temporal-State. They will be very much alive. This should be conveyed to the bereaving.

The dying process, therefore, is a Transition from the Temporal State to the Eternal Stage. Humans are, thereby, "restored to life."

With Transition, all faculties are also restored, including full access to all personal and ancestral memories.

Humans in the Sequential State will have access to all of the memories of their ancestors [78], from their parents back to Adam, the first man. They will also have access to all of the knowledge that God has chosen to share with them.

---

[77] Decomposition of cells
[78] This refers only to events that occurred prior to the conception of the next in line.

Section 17 – Death-Transition

Restored to life with atemporal bodies, humans will experience perfect, unaided communications between their souls and (spiritual) bodies, since both souls and bodies will exist in the Sequential State.

1 Corinthians 15:42-44 (TNIV) "So will it be with the resurrection of the dead. The body is sown perishable; it is raised imperishable; it is sown in dishonor; it is raised in glory; it is sown in weakness; it is raised in power; it is sown a natural body; it is raised a spiritual body."

This last Section is probably the most important. You have or probably will be with people, and their relatives, when they "die." They will be going through a process of death as described. They will not truly be physically dead until communications between the Temporal-State body and their soul can no longer be restored. (Corruption, or destruction, of the neural networks of the Temporal-State brain is the only positive proof of physical death.)

When the heart stops and breathing ceases, the Spirit with the soul of the dying person may soon leave the physical body. This is because all nerve cells stop communicating with the brain and therefore the neurons of the brain's cerebral cortex stop communicating with the Particle.[79] They may be able to see and hear everyone around them even if they were blind and deaf in life. If they died in great pain, they will now feel

---

[79] Many people have experimented with sensory-deprivation tanks and have archived Out-of-Body experiences.

Section 17 – Death-Transition

wonderful.

You may immediately be able to call them back to life, but you should attempt to do so only if you firmly believe they should not have died at that moment. Brain cells begin to die within three to seven minutes (depending on the temperature of the body), after which resuscitation may not be possible. [80]

We can't know how long the dying person will remain in the Out-of-Body state and stay in the area where their physical body died, which seems to justify the importance of (and reason for) the vigil.

Some have said that hearing is the last sense to remain as one dies. Others who have actually "died" and were resuscitated say that hearing stops first, then eyesight, then all other senses. However, the dying person's hearing is immediately restored as they experience the Out-of-Body state as explained by the Theory.

Yes, death is explained by the Atemporal Particle Theory. "**Death has lost its definition**," is the subtitle of my novel, Bridge. People who have been declared medically dead, with no brain activity, have been resuscitated. Refer back to the Pam Reynold's report by Dr. Sabom in which all of the blood had been drained from Pam's brain causing it to be incapable of thoughts.

---

[80] Resuscitation has been achieved 90 minutes after cardiac arrest even without using hypothermia to temporarily reduce the oxygen and metabolic requirements.

Section 17 – Death-Transition

I read once that a church member asked their pastor, "What happens when I die?" The pastor's answer was, "Well, your soul goes to heaven and you pass away." The church member then asked, "Pass away where?" The pastor had no answer.

Based on the Atemporal Particle Theory, if that same church member would then come to me and ask, "Really, what happens when I die?" I would say, "The short answer is, you don't. The longer answer is, you, in your Sequential State incorporeal body, go on living indefinitely. Where? You had better find a new pastor to ask."

It is reported by some who have had near-death experiences, in which immediately after their incorporeal body left their corporeal body, they saw their life flashing passed their eyes as in a slide presentation or a video. What actually happened was that their incorporeal body's brain was able to recall instantly everything that happened in their lives, sometimes with a focus on things they wish they hadn't done.

On a personal note, my father died with my sister, Ruth, at his bedside. As he transitioned, he apparently saw my mother, also named Ruth, enter the room. Mom had transitioned many years earlier. My father said, "Oh, now there are two Ruths in the room," after which his physical body died, and he joined my mother.

Section 18
Conclusions

Conclusion regarding Dr. Francis Crick's "Astonishing Hypothesis"

In some parts of Dr. Crick's book, *The Astonishing Hypothesis*, he seems to express his amazement about the construction of the brain when compared to that of a digital computer. He also recognizes the relative difference in operating speed (many millions of operations per second for computers compared to "in the region of only 100 spikes per second" for a neuron) and the problem this presents to how man sees, thinks and recalls memories.

Dr. Crick says, "A brain does not look even a little bit like a general purpose computer. Different parts of the brain, even different parts of the neocortex, specialize, at least to some extent, in handling different sorts of information. <u>Most memory appears to be stored in the very same locations that carry out current operations.</u>"

Hawkins tells his readers that the neocortex is about 2 millimeters (.079 inches) thick and, stretched flat, is

Section 18 – Conclusions

roughly the size of a large dinner napkin and that no one knows precisely how many cells it contains. Some anatomists have estimated that the human neocortex contains around 30 billion neurons, but Jeff Hawkins believes "those 30 billion neurons are you." He adds, "After twenty-five years of thinking about brains, I still find this fact astounding."

Hawkins agrees again with Crick. In *On Intelligence* he says, ". . . the brain's architecture has a great deal to tell us how the brain really works and why it is fundamentally different from a computer."

The *Atemporal Particle Theory* explains the apparent speed of the brain (with memory recall many times faster than the digital computer), and it suggests how the specialized parts of the brain are all interconnected (through the atemporal Particle [81]), and also why most memory appears to be stored in the very same locations that carry out current operations (memories are efficiently stored in the Particle and when recalled are returned to the same network of neurons that transferred them).

> **"The sensual and spiritual are linked together by a mysterious bond, sensed by our emotions, though hidden from our eyes." - Karl Wilhelm Von Humboldt (1767-1835)**

---

[81] The Atemporal Particle Theory became a theory when its hypothesis provided multiple solutions to related heretofore unanswered questions through many observations and a key aspect was found to be testable with expected scientific results. (Some scientists will disagree)

Section 18 – Conclusions

Dr. Crick's explanation for the unusual structure of the brain is as follows: "While a computer has been deliberately designed by engineers, the brain has evolved by natural selection over many, many generations of animals. This tends to produce a radically different style of design."

I am a design engineer, not a physicist and biochemist as Dr. Crick was. Here I can speak with greater authority. If the brain functions in a far, far superior manner than a computer, as we know it does, it is only logical to believe its design is superior, not just radically different. I attribute the superior design to a superior Designer. Dr. Crick attributes the "radically different style of design" to "natural selection over many, many generations of animals." He says this but yet cannot explain the evolution of the smallest component of the brain, the neuron.

Dr. Crick calls on the scientific community to prove his hypothesis. So do I call on the scientific community to test my Theory. And, borrowing his words, "If the scientific facts (gathered by the scientific community) are sufficiently striking and well established, and if they support" The Atemporal Particle Theory, then man will have proven the existence of the human soul, **a discovery that will surpass in importance the discovery of the structure of DNA**. At the same time, it will reinforce "the beliefs of billions of human beings alive today."

Section 18 – Conclusions

Conclusion regarding Jeff Hawkin's *On Intelligence*.

Hawkins says, "Intelligence is measured by the capacity to remember and (compare) patterns in the world, including language, mathematics, physical properties of objects and social situations. Your brain receives patterns from the outside world, stores them as memories, and makes (comparisons) by combining what it has seen before and what is happening now." He adds in his own words, "The recalled memory is compared with the sensory input stream."

Hawkins could have written and supported The Atemporal Particle Theory better than I. He seems to be missing only one important component - the Particle - without which his entire theory is unworkable.

The Atemporal Particle Theory Summary

There are three states of existence, the Temporal State, the Sequential State and the Concurrent State.

In the Temporal State, beings are limited by time and sequence. In the Sequential State, beings are limited only by sequence. In the Concurrent State, God is not limited by time nor sequence.

We are body/soul/spirit (triune) beings. The physical body is in the Temporal State ("Body"). The Spirit is in the Sequential State and it is identical to the Body except that it does not age past maturity. The Spirit, being atemporal and without mass, occupies the same space

Section 18 – Conclusions

as the Body. The soul is in the Sequential State and has been herein referred to as the Particle.

The Particle is the source of all information regarding the Body and Spirit, with the exception of the physical traits stored in DNA. The Particle is the repository of all memories including ancestral memories. [82]

At conception, the Particle of each parent, as well as their DNA, combine. The Particle and DNA are therefore unique to each individual as is the Body and Spirit which develop as a result of conception.

The Particle and DNA are at the functional center of every cell in the Body, [83] with the exception of red blood cells. All cells receive instructions from the Particle, but brain cells (Neurons) share information with the Particle.

DNA is wrapped 1.65 turns around bead-like histones forming nucleosomes. The assembly is twisted tightly forming Solenoid Chromatin Fibers.[84]

---

[82] An ancestral memory skill surfaces when an individual of an early age is in possession of an extraordinary aptitude for certain mental activity, which showed itself at so early a period as to exclude the notion that it could have been acquired by the experience of the individual.

[83] Being atemporal and without mass, the same Particle can be at the functional center of each cell.

[84] Preceding mitosis, the Solenoid Chromatin Fibers are further coiled, folded and packed into a chromosome.

Section 18 – Conclusions

At rest, the Neuron's inner membrane wall is about 70 millivolts negative, relative to the outer wall. When the Neuron receives information to be transferred to the Particle, an Action Potential occurs, reversing the polarization of the cell's membrane to +40 millivolts.

The Action Potential thereby creates a current in the Solenoid Chromatin Fiber within the Neuron which produces a very high frequency electromagnetic Field Signal enveloping the Particle.[85]

Neurons and other cells receive information from the Particle which produces a corresponding Field Signal enveloping the Solenoid Chromatin Fiber of the cells.

Neurons communicate constantly with the Particle as they receive incomprehensible information at random from billions of other Neurons and non-brain cells of the Central Nervous System. The Particle instantly returns the information as pulsed, comprehensible percepts, making consciousness possible.

External electromagnetic radiation can interfere with the

---

[85] One explanation for how this occurs follows: The capacitance of the neuron body, having at rest a positive wall and a corresponding negative wall, is coupled with the parasitic capacitance of the solenoid chromatin fibers within the nucleus of the neuron. The rapid reversal of the charge on the walls of the neuron during an action potential causes a reversal of the charges on the parasitic capacitors of the solenoid chromatin fibers, which are effectively in parallel with the solenoids, and thereby a current is produced within each of the solenoids which in turn generates an electromagnetic field carrying the information from the neuron to the Particle.

Section 18 – Conclusions

Field Signals and therefore, the information transferred to cells and from Neurons may be affected, causing errors in communications especially significant during mitosis. Communications between the Neurons of the Body and the Particle are slowed by the physical limitations [86] of the Body's brain especially as the Body ages [87]. On the other hand, communications between the neurons of the Spirit and the Particle are not slowed by any physical limitations.

If the Particle is not able, for any reason, to be in constant communications with certain sections of an individual's physical brain (such as the cerebral cortex, frontal or temporal lobes), the individual's Spirit may entirely leave the Body as it does in death or in an out-of-body experience (OBE).

When the Spirit leaves the Body, the individual will, in the Sequential State, experience consciousness with full mental faculties [88] and no time between events (limited only by sequence).

If the Spirit leaves the Body, but the Particle can once again establish constant communications with those key

---

[86] The Neurons of the brain transfer information through an action potential (which in the Body's brain takes about 5ms) inducing a current in the Solenoid Chromatin Fiber (requiring time in the Body's brain due to inductive reactance) which in turn generates an electromagnetic field enveloping the Particle. The Particle then immediately returns the information through the same process to the Neurons. The Spirit's brain processes information in the Sequential State with no time involved.
[87] Resulting in damages to the Solenoid Chromatin Fiber DNA.
[88] Including sight and hearing even though the Body of the individual does not have those faculties.

Section 18 – Conclusions

sections of the brain, resuscitation is possible (or return to the Body in the case of an OBE).

If the Particle is not able, for any reason,[89] to be in constant communications with only a small part of an individual's brain, the neurons of the individual's Spirit may function in place of those neurons that are not communicating and they will function in the Sequential State, again with no time between events.

Also, if the Particle is not able, for any reason, to transfer information to non-neuron cells, the cells of the individual's Spirit may function for those cells where possible.

The Atemporal Particle Theory explains the following

Out-of-body Experiences (OBE)
Individuals may enter the Sequential State through sensory deprivation, reducing communications between the Particle and the brain. It is not known how long an individual can remain in this state.

Near-death Experiences (NDE)
Some individuals enter the Sequential State when the brain receives too little oxygen to communicate with the Particle, and when sufficient oxygen is restored, they are resuscitated. Some report having been able to recall almost instantly their entire lives.

---

[89] Such as disease, damage or interference with the Field Signal in a section of the brain.

Section 18 – Conclusions

Hyperthymesia
Hyperthymesia is the condition of possessing an extremely detailed autobiographical memory. The neural-network of the Body's brain,[90] involved in long-term memory recall, may have been injured (or never developed properly from birth) and is unable to communicate with the Particle. The matching neural-network of the Spirit's brain apparently functions instead in the Sequential State.

Phantom Limb Syndrome
Phantom limb syndrome is the sensation that an amputated or missing limb is still attached to the body and is moving appropriately with other body parts. Approximately 60 to 80% of individuals with an amputation experience phantom sensations in their amputated limb. These individuals are sensing the existing limbs of the Spirit.

Savant Syndrome
Savant syndrome is a condition in which a person demonstrates abilities far in excess of what would be considered normal. People with savant syndrome may have neurodevelopmental disorders or brain injuries. Those parts of the brain producing the unusual abilities[91] have been substituted by the matching neural-network of the Spirit's brain and are functioning in the Sequential State.

---

[90] Perhaps the hippocampus and the prefrontal cortex.
[91] Artistic brilliance, mathematical mastery, photographic memory.

Section 18 – Conclusions

Testing the Atemporal Particle Theory

It should be possible to construct a small model of a solenoid suspended in a three-walled sphere simulating a solenoid chromatin fiber within the nucleus of a neuron as proposed by the following graphic.

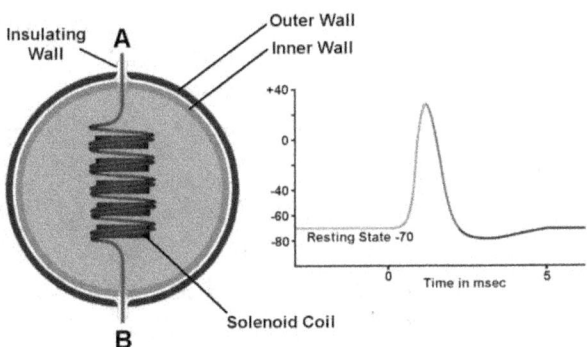

At rest, the inner wall will be maintained at -70 millivolts relative to the outer wall. An oscilloscope will be attached to terminals A and B. The voltage applied will then be reversed as shown in the graph on the right and returned to the resting state in 5 milliseconds. An oscillating voltage should appear on the oscilloscope demonstrating that it is possible that a neuron, during an action potential, will generate a voltage in the solenoid chromatin fibers within the walls of the neuron, and that they will in turn generate an electromagnetic field, assuming, of course, that the solenoid and sphere used in the test perform like the solenoid chromatin fiber and neuron as suggested.

During my studies of savants and those individuals with hyperthymesia, I recognized that they probably

Section 18 – Conclusions

presented an opportunity to demonstrate the Theory, because it appeared that the brains of those individuals function far faster than possible for normal individuals. Part of their brains must function in the Sequential State.

Scientist Allan Snyder [92]

Snyder is interested in understanding savants, how the savant brain perceives and interprets the world, the neurological and subjective correlates of the savant brain, and how to activate or at least promote savant level brain functions in non-autistic, healthy individuals. In savants, says Snyder, the top layer of mental processing (conceptual thinking, making logical deductions) is somehow deactivated. His working hypothesis is that once this layer is inactivated, one can access a startling capacity for recalling the most minute detail or for performing lightning-quick calculations.

Dr. Bruce Miller [93]

Dr. Miller, working with patients having frontotemporal dementia, a degenerative brain disease that strikes people in their fifties and sixties, has seen these patients spontaneously develop both interest and skill in art and music. Patients with damage in this area can't name what they're looking at, but they can often paint it beautifully. Miller has also seen physiological similarities in the brains of autistic savants and patients with frontotemporal dementia. When he performed brain-imaging studies on an autistic savant artist who started

---

[92] Director of the Centre for the Mind at the University of Sydney.
[93] A neurologist at the University of California at San Francisco.

Section 18 – Conclusions

drawing horses at 18 months, he saw abnormalities similar to those of artists with frontotemporal dementia: decreased blood flow (therefore oxygen) and slowed neuronal firing in the left temporal lobe.

Final Comment

To the scientist who believes in God, His Son, and the Holy Spirit I say the following: I have read some of your books defending your belief in a Creator. I have attended some of the conferences in which you called Darwin's theory of natural selection preposterous, when it comes to the evolution of the cell. I listened to some of your talks about the "Fingerprints of God" on the creation of the human cell, but when I asked about the human soul you answered, "I don't know how the soul figures into all of this." Did God forget to tell you, or did you forget to ask? Once you know, the world needs to hear from you.

In summary, we have a physical body which ages and, concurrently, a spiritual body which does not. Both are in communications with the soul, a spiritual component of the human where all memory resides. Each soul is the union of the souls of the parents. Soul-brain communications make consciousness possible. If the soul can no longer communicate with the physical body, the spiritual body/soul unity (again defining "Spirit") leaves the physical body, and, if resuscitation is impossible, arrives in the presence of God. However, be aware, the Spirit may remain to hear and see what is occurring around the physical body. Growth and healing processes are directed by the soul, as are all cellular

Section 18 – Conclusions

functions. Miraculous healing requires faith on the part of the recipient, due to free will, but the normal healing processes take place, except that they may be substantially accelerated. The warmth recipients often feel is due to the accelerated metabolic processes.

Section 19
Christian Layperson's Addendum

Genesis 1:26 (NIV)
Then (Jesus) said, "Let us make mankind in our image, in our likeness, so that they may rule over the fish in the sea and the birds in the sky, over the livestock and all the wild animals, and over all the creatures that move along the ground."

Genesis 2:7 (NIV)
Then the Lord God formed a man from the dust of the ground and breathed into his nostrils the breath of life, and the man became a living being.

1 Thessalonians 5:23 (TNIV)
"May God himself, the God of peace, sanctify you through and through. May your whole <u>spirit</u>, <u>soul</u> and <u>body</u> be kept blameless at the coming of our Lord Jesus Christ."

Set aside your current understanding of who and what you are. Yes, you have a spirit, soul and body. You were created a triune (three part) being.

Section 19 – Layperson's Addendum

As you read this, your physical body ("Body") and your spiritual body ("Spirit") are coincident; they occupy the same space at the same time. Your Body and Spirit are together with every movement you make. They are inseparable, except as I will explain.

Your soul ("Soul") is at the functional center of every cell in your Body and Spirit. It is your Soul that is the repository of all your memories and those of your ancestors.

> "The fact is that the brain doesn't store memories, and can't store memories," Dr. Michael Egnor, Professor and Vice-Chairman, Department of Neurological Surgery, Director, Pediatric Neurosurgery, Stony Brook Medicine University Physicians.

> "The brain itself does not store (long term) memories at all, but rather retrieves them from an external store. Memories are recovered atemporally," Nicholas H.E. Prince, Mathematical Physicist.

It is your Soul that directs the biological functions of all of your cells, not the DNA contained within.

> "Life is the control of the chemical (DNA), not the control by the chemical." Dr. David Wilcox, Professor of Biology at Eastern University in St. Davids, Pennsylvania.

Section 19 – Layperson's Addendum

Your Soul is **not** you. It is a functional part of you, but it is not you.

Your Soul is communicating continuously with every cell in your Body and Spirit. With the cells in your central nervous system, there is two-way communications.

> "The process is generally treated as though information flows in a single direction, but information in the cortex always flows in the opposite direction as well," Jeff Hawkins, Executive Director and Chairman of the Redwood Neuroscience Institute.

If your Body should lose a limb, your Spirit's limb is still there (explaining the phantom limb, phantom pain phenomena).

> "Approximately 60 to 80% of individuals with an amputation experience phantom sensations in their amputated limb," Study Guide for Medical Surgical Nursing Care.

> "Although phantom pain occurs most often in people who've had an arm or leg removed, the disorder may also occur after surgeries to remove other body parts," Mayo Clinic.

If a part of your Body's brain is damaged or diseased, or some part is caused by some other means to stop communicating with your Soul, the neurons of your Spirit may function in their place. However, your Spirit is not limited by time, as is your Body, and the results may

Section 19 – Layperson's Addendum

be far in excess of what would be considered normal. This is also true for some individuals whose brains have never developed properly.

> "In savants the top layer of mental processing is somehow deactivated. Once this layer is inactivated, one can access a startling capacity for recalling the most minute detail or for performing lightning-quick calculations," Dr. Allan Snyder, Director of the Centre for the Mind at the University of Sydney.
>
> Dr. Bruce Miller, a neurologist at the University of California at San Francisco, working with patients having a degenerative brain disease that strikes people in their fifties and sixties, has seen these patients spontaneously develop both interest and skill in art and music.

Now, if for any reason, the major control centers in your Body's brain stop communicating with the Soul altogether, your Spirit and Soul may separate from your Body. This happens when your Body's brain can no longer function or when, through sensory deprivation, your Body's brain has no information to transfer to your Soul.

Considering the latter first, through sensory deprivation, an out-of-body experience may occur. You will continue to function as a Spirit with no time restrictions. You will be able to see and hear what is going on around you, but those who remain in their bodies will generally not be able to see or hear you. You will be able to move

Section 19 – Layperson's Addendum

through physical objects and travel great distances in an instant. You can, at will, return to your Body at any time.

If for any cause, the major control centers in your Body's brain stop communicating with the Soul altogether, because your Body's brain physically cannot function, your Spirit and Soul will separate from your Body. You will be fully conscious, you will be able to see and hear what is going on around you, but others will generally not be able to see or hear you. You will feel calm and well. You will be able to move through physical objects and travel great distances in an instant. You may willingly proceed toward a distant place where you may also meet others who have previously gone before you. This is the process of dying.

On the other hand, if the physical cause for your Body's brain to stop communicating with your Soul is reversed before cellular corruption takes place, you may be resuscitated and return to your Body. If it is not, you will find yourself, as a Spirit with a soul, in the presence of God.

> "Jesus called out with a loud voice, 'Father, into your hands I commit my spirit.' When he had said this, he breathed his last," Luke 23:46

Section 20
Bibliography

Barbour, Julian, *The End of Time*, The Next Revolution in Physics, 1999
Behe, Michael J., *Darwin's Black Box*, The Biochemical Challenge to Evolution, 1996
Beisswenger, Kai, *Zeitpuzzle*, a novel, 2002
Beveridge, W.I.B., *The Art of Scientific Investigation*, An Entirely Fresh Approach to the Intellectual Adventure of Scientific Research, 1950
Brown, Warren S. and Murphy, Nancy and Malony, H. Newton, *Whatever Happened to the Soul*, Scientific and Theological Portraits of Human Nature, 1998
Chopra, Deepak, M.D., *Ageless Body, Timeless Mind*, The Quantum Alternative to Growing Old, 1993
Crick, H.C. Francis, *The Astonishing Hypothesis*, The Scientific Search for the Soul, 1995
Davies, Paul, *About Time*, Einstein's Unfinished Revolution, 1995
Davies, Paul, *The Mind of God*, The Scientific Basis for a Rational World, 1993
Edwards, Robert and Steptoe, Patrick, *A Matter of Life*, The Story of a Medical Breakthrough, 1980
Einstein, Albert, *Out of My Later Years*, 1950, 1990 Edition
Faid, Robert W., *A Scientific Approach to Christianity*, New Evidence Supports the Bible!, 1982
Fischbach, Gerald D., *How Neurons Communicate*, Scientific American, September, 1992
Hawkins, Jeff with Sandra Blakeslee, *On Intelligence*, How a New Understanding of the Brain Will Lead to the Creation of Truly Intelligent Machines, 2004
Landau, David, *Death is Not Always the Winner*, a novel, 2001
Levine, Arnold J., *Viruses*, 1992
McInerny, Ralph M. Ph.D., *Miracles*, A Catholic View, 1986

McTaggart, Lynne, *The Field*, The Quest for the Secret Force of the Universe, 2002
Miller, Joseph D., Ph.D. *The Prospects for a Quantum Neurobiology*, 1997
Mills, Dr. Randell L., *The Grand Unified Theory of Classical Quantum Mechanics*.
Milton, Richard, *Shattering the Myths of Darwinism*, 1997
Moody, Raymond A., M.D., *The Light Beyond*, New Explorations by the Author of *Life After Death*, 1988
Pearsall, Paul, Ph.D., *The Heart's Code*, Tapping the Wisdom and Power of Our Heart Energy, 1998
Prince, Nicholas H.E., *Are Memories Really Stored in the Brain?*, 2003
Schein, Elyse and Bernstein, Paula, *Identical Strangers*, A Memoir of Twins Separated and Reunited, 2008
Schwartz, Gary E., Ph.D., *The Afterlife Experiments*, Breakthrough Scientific Evidence of Life After Death, 2001
Ratzinger, Joseph Cardinal, Interdicasterial Commission for the *Catechism of the Catholic Church*, 1994
Rensberger, Boyce, *Life Itself*, Exploring the Realm of the Living Cell, 1995
Ross, Hugh, Ph.D., *Creation and Time*, A Biblical and Scientific Perspective on the Creation-Date Controversy, 1994
Ross, Hugh, Ph.D., *The Creator and the Cosmos*, How the Greatest Scientific Discoveries of the Century Reveal God, 1993
Sabom, Michael, M.D., *Light & Death*, One Doctor's Fascinating Account of Near Death Experiences, 1998
Schroeder, Gerald L., *Genesis and the Big Bang*, The Discovery of Harmony Between Modern Science and the Bible, 1992
Schwartz, Gary E., Ph.D. and Chopra, Deepak, M.D., *Science and Soul*, The Survival of Consciousness After Death, a tape dialog, 2001

Sheldrake, Rupert, Ph.D., *A New Science of Life,* The Hypothesis of Morphic Resonance, 1995
Siegel, Bernie, M.D., *Love, Medicine and Miracles,* Lessons Learned About Self-Healing from a Surgeon's Experience with Exceptional Patients, 1986
Thorne, Kip S., *Black Holes & Time Warps*, Einstein's Outrageous Legacy, 1994
Ward, Keith, *In Defense of the Soul*, 1998
Wigglesworth, Smith, *Healing*, Experience God's Miracles, 1982

Other references can be found in the body of the text and the following Index.

Section 21
Index

As you will see, this Index was specially prepared for your research.

A
A New Science of Life, 146
action potentials, 82, 83, 110
Adequate sleep, 114
aging, 112, 113
Aging, 112, 113
Allan Snyder, 136
ancestral, 28, 122, 130
ancestral memories, 28, 122, 130
Are Memories Really Stored in the Brain, 64, 145
association areas, 86
Astonishing Hypothesis, 1, 3, 8, 19, 22, 23, 126, 144
atemporal, 10, 27, 28, 110, 120, 123, 127, 129, 130
atemporally, 28, 64
atheist, 28
atheistic, 29
Atwater, P.H.M, 118
autobiographical memory, 134
axon, 82, 83

B
Ball, John, 70
benign childhood epilepsy, 86
Bible, 116, 119, 144, 145
biological, 22, 25, 68, 112
Body, 42, 76

Body/Spirit unity, 42, 76
brain, 23, 24, 29, 53, 54, 55, 63, 64, 65, 66, 67, 68, 69, 70, 83, 84, 85, 87, 88, 115, 117, 123, 126, 127, 128, 129, 130, 131, 132, 133, 134, 136, 137

C

cancer, 101, 105, 106, 107
cancer cell, 106, 107
cell, 42, 76
cell division, 100, 101, 105, 106
cell-specific instructions, 105
cellular phone, 115
central nervous system, 42, 76
centrosome, 106
chromosome, 84, 106, 108, 130
chromosomes, 105
coded genetic traits, 84
Coincident, 40
communicating, 42, 76
communicating continuously, 42, 76
comparisons, 67, 129
comprehensible, 87, 88, 131
computer, 67, 68, 126, 127, 128
conception, 101, 122, 130
Concurrent, 129
Concurrent State, 129
consciousness, 25, 29, 47, 66, 110, 131, 132, 137
control and control process, 69
controlled, 54
Corresponding, 105, 106, 109
corruption, 122
cortex, 27, 65, 66, 86, 132, 134
cosmos, 22, 25
creation, 27, 111, 137

Creation, 23, 144, 145
Creator, 23, 137, 145
Crick, Francis, 1, 22, 23, 24, 25, 29, 70, 87, 126, 127, 128

D

Darwin, 137, 144
death, 110, 116, 118, 119, 120, 122, 123, 132, 133
degenerative brain disease, 55, 136
dendrites, 82, 85
determinants of longevity, 113
digital computer, 67, 126, 127
disturbance, 84, 85, 110
division limits, 105
DNA, 22, 23, 84, 87, 88, 101, 105, 106, 107, 108, 113, 114, 128, 130, 132
DNA polymerase molecule, 108
Dr. Bruce Miller, 136
dying process, 121, 122

E

electromagnetic field, 115, 131, 132, 135
electromagnetic fields, 115
Electromagnetic Radiation, 109
epithelial cells, 109, 110
Erlykin and Wolfendale, 114
Eternal Stage, 122
evolved, 22, 128
existence, 22, 23, 30, 115, 116, 128, 129
exonuclease, 108
extemporaneous discussions, 70
extemporaneous speech, 67

F

Field, 66, 84, 85, 86, 88, 105, 106, 108, 109, 114, 121, 131, 132, 133, 145
fingerprints of God, 137
Fischbach, 82, 85, 144
frontotemporal dementia, 55, 136

G

Gilman, Robert, 63
Glover, David, 106
God, 23, 27, 28, 110, 115, 119, 122, 129, 137, 144, 145, 146
Grolier Multimedia Encyclopedia, 113
growth, 104, 105, 106

H

Hardin, Garrett, 23
Hawking, Stephen, 25
Hawkins, Jeff, 1, 24, 26, 27, 28, 29, 65, 66, 67, 68, 69, 83, 86, 87, 126, 127, 129, 144
Hayflick, Leonard, 112, 113
healing, 104, 105, 110, 137
human, 20, 22, 23, 68, 87, 99, 100, 102, 114, 127, 128, 137
Humboldt, Karl, 127
Hyperthymesia, 134
hypothetical statements, 65, 119

I

intelligence, 24, 25, 29
intuition, 27

L

Lazarus, 119
life, 114, 118, 119, 120, 122, 123, 124
life-limiting radiation, 114
Likeness, 104, 111, 114
love, 53
Love, Medicine and Miracles, 121, 146
low-level electromagnetic radiation, 115

M

memories, 27, 28, 63, 64, 65, 66, 67, 122, 126, 127, 129, 130
memory, 27, 64, 65, 67, 68, 70, 87, 114, 115, 126, 127, 129, 134, 137
Memory, 63, 64, 69
microtubules, 106
Miller, 55, 56, 136
mind, 23, 88, 110, 119
mitosis, 84, 106, 108, 130, 132
mother, 53

N

natural selection, 22, 112, 128, 137
Near-death Experiences, 133
Near-death Experiences (NDE), 133
Near-Death Stage, 121
neocortex, 27, 65, 69, 126
nerve cells, 123
nervous system, 110
neuroscience, 24
neural-network, 134
neurocortex, 65
neurological, 88, 136

neuron, 67, 69, 70, 82, 83, 84, 85, 86, 88, 110, 120, 121, 126, 127, 128, 131, 133, 135
neuro-network, 65, 67
neurons, 20, 27, 42, 66, 67, 68, 70, 76, 82, 84, 85, 86, 87, 110, 114, 123, 127, 132
neurotransmitter, 83
New Scientist, 64
non-neuron cells, 133
nucleosome assembly, 85
nucleosomes, 84, 130
nucleotides, 108
nucleus, 84, 85, 88, 109, 131, 135

O

On Intellegence, 29, 127, 129
On Intelligence, 1, 26
one hundred-step rule, 68
operating speed, 126
oscilloscope, 135
Out-of-Body experience, 117
Out-of-body Experiences, 133
Out-of-body Experiences (OBE), 133
Out-of-Body Stage, 121

P

Parallata, 64, 65, 66, 69, 86, 87
Parallatum, 65, 86
parallel computer, 67, 68
parents, 53
Particle, 1, 3, 23, 27, 40, 64, 65, 66, 67, 69, 70, 84, 85, 86, 87, 88, 104, 105, 106, 107, 108, 109, 110, 113, 114, 120, 121, 123, 127, 128, 129, 130, 131, 132, 133, 134, 135

Particle,, 64, 66, 84, 88, 104, 105, 106, 114, 130, 131, 133
Pearsall, Paul, 63, 145
percepts, 69, 131
pericentriolar material, 106
Phantom Limb Syndrome, 134
physical regeneration, 104
Plato, 26, 27, 29
polymerase, 108, 113
postsynaptic membrane, 83
postsynaptic neuron, 83
power lines, 115
predictions, 67
prefrontal cortex, 134
primary auditory area, 86
primary auditory area - A1, 86
primary sensory areas, 86
primary somatosensory region, 86
primary somatosensory region - S1, 86
primary visual area, 86
primary visual area - V1, 86
Prince, Nicholas, 64, 145
processes of aging, 113
pyramidal cell, 69

R

radiation, 84, 85, 114, 131
Reacholi, Michael, 114
reality, 26, 27, 28, 29
religion, 28
REM sleep, 110, 114
removed, 53, 54
repository, 130
reproduction, 113, 114

resurrection, 119, 123
resuscitated, 124, 133
resuscitation, 121, 124, 133, 137
retention, 63
returned from death's door, 116
Reynolds, Pam, 117, 118
RNA, 113
Ross, Hugh, 114, 145

S

Sabom, Michael, 116, 117, 145
Savant syndrome, 134
Scientific American, 82, 106, 144
scientist, 23, 28, 116, 137
scientists, 24, 65, 70, 127
sequence of patterns, 66
Sequential, 64, 66, 86, 88, 111, 120, 121, 122, 123, 129, 132, 133, 134, 136
Sequential State, 120, 122, 123, 129, 132, 133, 134, 136
sequentially, 65, 66, 111
Sequential-State body/soul unity, 121, 122
Sheldrake, Rupert, 146
Siegel, Bernie, 121, 146
sleep, 110, 114, 119
Snyder (Alan) 136
Solenoid Chromatin Fiber, 84, 85, 105, 130, 131, 132
Somehow, 108
soul, 23, 25, 26, 27, 64, 67, 70, 87, 88, 111, 119, 120, 121, 122, 123, 128, 129, 137
Soul, 137, 144, 145, 146
Spinal reflexes, 70
spirit, 88, 119, 120, 122, 123, 129, 137
Spirit, 121, 122, 130, 132, 133, 134, 137

Standstill, 117
supernovae, 114
synapses, 67, 69, 82, 83, 85

T

temporal, 54, 55, 56, 64, 110, 120, 132, 137
Temporal, 64, 85, 87, 88, 113, 120, 121, 122, 123, 129
Temporal State, 113, 120, 122, 129
The Atemporal Particle Theory, 3, 23, 30, 40, 84, 127, 128, 129, 133, 135
The Heart's Code, 63, 145
Thinking Solutions, 70
Tillis, Ray, 64
transition, 116
Transition, 3, 116, 122
transplant recipients, 63
Truly Astonishing Hypothesis, 4, 8, 19, 22, 84, 128, 129
two way communications, 42

U

unity, 66, 88, 121, 122, 123, 137
universe, 28

V

Vela supernova, 114
vision, 88

W

WHO's Electromagnetic Fields Project, 115

Z

zygote, 22, 23

End Notes

[i] Dr. Crick is not the only one puzzled by this problem. Sapien Labs is a not-for-profit organization established in 2016 with a mission to enable deeper understanding of the human mind. They also recognize the Binding Problem. Read #4.

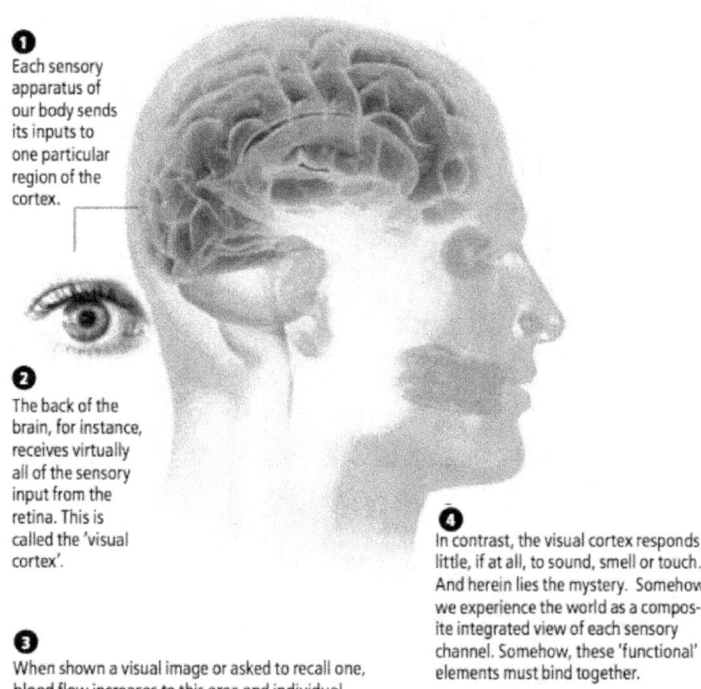

[THE BINDING PROBLEM]
HOW DOES IT ALL COME TOGETHER?

❶ Each sensory apparatus of our body sends its inputs to one particular region of the cortex.

❷ The back of the brain, for instance, receives virtually all of the sensory input from the retina. This is called the 'visual cortex'.

❸ When shown a visual image or asked to recall one, blood flow increases to this area and individual neurons in these regions fire more strongly.

❹ In contrast, the visual cortex responds little, if at all, to sound, smell or touch. And herein lies the mystery. Somehow, we experience the world as a composite integrated view of each sensory channel. Somehow, these 'functional' elements must bind together.

sapienlabs.org

[ii] "Experiments by the early part of the 20th century had revealed that both light and electrons behave as waves in certain instances and as particles in others. This was unanticipated from preconceptions about the

nature of light and the electron. Early 20th century theoreticians proclaimed that light and atomic particles have a 'wave-particle duality' that was unlike anything in our common-day experience. The wave-particle duality is the central mystery of the presently accepted atomic model, quantum mechanics, the one to which all other mysteries could ultimately be reduced." Dr. Randell L. Mills, in his book, *The Grand Unified Theory of Classical Quantum Mechanics*.

Is Dr. Mills suggesting that all matter has a dual nature, one physical and one atemporal and that understanding the dual nature of all matter will lead to . . . ?

"There must be a point in this reductionist program where molecular biology enters the domain of quantum physics, a point at which classical, Newtonian, deterministic theory (the usually unacknowledged underpinning of modern biology) must give way to quantum mechanical interpretation. Nowhere will this be seen more clearly than in attempts to understand the mechanism and function of the central nervous system and the diseases to which it is prone. Already, quantum mechanical considerations are necessary in the modeling of neurotransmitter receptor structure (Pardo et al., 1996). It is possible that some of the "hard" problems in neurobiology, such as the nature, origin, and development of conscious experience will require a quantum perspective. The inevitable intersection of neurobiology with quantum mechanics may lead to a twin understanding of cognition and the role of the observer in quantum mechanics. That knowledge in turn

will have profound implications for psychiatric medicine, the definition of 'human', and perhaps the interpretation of physical reality itself. But to evaluate such possibilities it is necessary to briefly consider both philosophical approaches to the nature of consciousness and the essentials of quantum mechanics. The difficulties in the interpretation of quantum mechanics (as opposed to its pragmatic application) may provide unique insight into some of the most difficult problems of neurobiology."

The Prospects for a Quantum Neurobiology by Joseph D. Miller Ph.D, Department of Pharmacology, Texas Tech University Health Sciences Center, 1997.

[iii] Douay-Rheims is one of the few interpretations that accurately translates what God said.

Here is a NIV translation: Moses said to God, "Suppose I go to the Israelites and say to them, 'The God of your fathers has sent me to you,' and they ask me, 'What is his name?' Then what shall I tell them?" God said to Moses, "I am who I am."

"I am who I am?"

Is that really what God said, I wondered? Is that really how God answered Moses? Is that a proper answer to the question, "Suppose I go to the Israelites and say to them, 'The God of your fathers has sent me to you,' and they ask me, 'What is his name?' Then what shall I tell them?" And God answers, "I am who I am?" Is that what God actually said? The Bible is the word of God, but did the translators get it right from the original Hebrew? Others aren't so sure either. Here's a list of

the various translations of God's answer in Exodus 3:14.

"I will exist which I will exist."	Modern Translation of Hebrew Bible
"I am who I am."	New International Version
"I will be what I will be."	New International Version (alternate)
"I am who I am."	New American Standard Bible
"I am who I am."	The Message
"I am who I am and what I am and I will be what I will be."	Amplified Bible
"I am who I am."	New Living Translation
"I will be what I will be."	New Living Translation (alternate)
"I am that I am."	King James Version
"I am who I am."	English Standard Version
"I am the eternal God."	Contemporary English Version
"I am who I am."	New King James Version
"I am who I am."	New Century Version
"I am that I am."	21st Century King James Version
"I am that I am."	American Standard Version
"I am that which I am."	Young's Literal Translation
"I am that I am."	Darby Translation
"I am who I am."	Holman Christian Standard Bible
"I am who I am."	New International Reader's Version

"I am who I am."	New International Version – UK
"I am who I am."	Today's New International Version
"I am who am."	New American Bible
"I am who am."	Douay-Rheims
"I am who I am."	Revised Standard
"I am he who is."	New Jerusalem Bible

So, of these translations, with all due respect, I ask, which would serve as a proper answer to Moses' question? Moses needed to know. After all, God had just said, "I am sending you to Pharaoh to bring my people the Israelites out of Egypt."

Moses was a shepherd. He was overwhelmed by the assignment. He needed a straight answer. If you were Moses, would "I am who I am?" be a good, reasonable, loving answer? Not for me. If I was asked by the publisher of this book, "Who should I tell the readers you are?" and I answered, "I am who I am." Would that have been a proper response?

How about the King James translation, "I am that I am?" Still doesn't answer the question as Moses needed it answered.

How about the Modern Translation of the Hebrew Bible, "I will exist which I will exist?" Nope. Or, "I will be what I will be?" Not that one either. Then there's, "I am who I am and what I am, and I will be what I will be?" Moses might have asked, "Would you say that again?" And also, "I am that which I am." Not any better.
That leaves us with three good answers. "I am the eternal God," "I am who am" and "I am he who is." Well,

"I am the eternal God" is certainly a great answer, but it is not a translation; it is a dynamic equivalent of what was said by God.

That leaves us with two interpretations which have the same meaning. God is the only being in existence who simply is. For him, everything is in the present.

Remember, now, God finished his answer with, "This is what you are to say to the Israelites: 'I AM has sent me to you.'" "I AM," therefore, provided Moses with a clear statement of who was sending him.

[iv] Douay-Rheims is one of the few interpretations that accurately translates what God said.

Here is a NIV translation: Moses said to God, "Suppose I go to the Israelites and say to them, 'The God of your fathers has sent me to you,' and they ask me, 'What is his name?' Then what shall I tell them?" God said to Moses, "I am who I am."

"I am who I am?"

Is that really what God said, I wondered? Is that really how God answered Moses? Is that a proper answer to the question, "Suppose I go to the Israelites and say to them, 'The God of your fathers has sent me to you,' and they ask me, 'What is his name?' Then what shall I tell them?" And God answers, "I am who I am?" Is that what God actually said? The Bible is the word of God, but did the translators get it right from the original Hebrew? Others aren't so sure either. Here's a list of the various translations of God's answer in Exodus 3:14.

"I will exist which I will exist."   Modern Translation of
                                      Hebrew Bible

"I am who I am."	New International Version
"I will be what I will be."	New International Version (alternate)
"I am who I am."	New American Standard Bible
"I am who I am."	The Message
"I am who I am and what I am and I will be what I will be."	Amplified Bible
"I am who I am."	New Living Translation
"I will be what I will be."	New Living Translation (alternate)
"I am that I am."	King James Version
"I am who I am."	English Standard Version
"I am the eternal God."	Contemporary English Version
"I am who I am."	New King James Version
"I am who I am."	New Century Version
"I am that I am."	21st Century King James Version
"I am that I am."	American Standard Version
"I am that which I am."	Young's Literal Translation
"I am that I am."	Darby Translation
"I am who I am."	Holman Christian Standard Bible
"I am who I am."	New International Reader's Version
"I am who I am."	New International Version – UK
"I am who I am."	Today's New International Version
"I am who am."	New American Bible

"I am who am."	Douay-Rheims Version
"I am who I am."	Revised Standard Version
"I am he who is."	New Jerusalem Bible

So, of these translations, with all due respect, I ask, which would serve as a proper answer to Moses' question? Moses needed to know. After all, God had just said, "I am sending you to Pharaoh to bring my people the Israelites out of Egypt."

Moses was a shepherd. He was overwhelmed by the assignment. He needed a straight answer. If you were Moses, would "I am who I am?" be a good, reasonable, loving answer? Not for me. If I was asked by the publisher of this book, "Who should I tell the readers you are?" and I answered, "I am who I am." Would that have been a proper response?

How about the King James translation, "I am that I am?" Still doesn't answer the question as Moses needed it answered.

How about the Modern Translation of the Hebrew Bible, "I will exist which I will exist?" Nope. Or, "I will be what I will be?" Not that one either. Then there's, "I am who I am and what I am, and I will be what I will be?" Moses might have asked, "Would you say that again?" And also, "I am that which I am." Not any better.

That leaves us with three good answers. "I am the eternal God," "I am who am" and "I am he who is." Well, "I am the eternal God" is certainly a great answer, but it is not a translation; it is a dynamic equivalent of what was said by God.

That leaves us with two interpretations which have the same meaning. God is the only being in existence who simply is. For him, everything is in the present.

Remember, now, God finished his answer with, "This is what you are to say to the Israelites: 'I AM has sent me to you.'" "I AM," therefore, provided Moses with a clear statement of who was sending him.

Here is the Hebrew text of Exodus 3:14

יד וַיֹּאמֶר אֱלֹהִים אֶל-מֹשֶׁה, אֶהְיֶה אֲשֶׁר אֶהְיֶה; וַיֹּאמֶר, כֹּה תֹאמַר לִבְנֵי יִשְׂרָאֵל, אֶהְיֶה, שְׁלָחַנִי אֲלֵיכֶם.

It reads from right to left. This is the subject phrase:

אֶהְיֶה אֲשֶׁר אֶהְיֶה

It translates literally, "Am who am." I conclude, "I am who am." is the correct translation. It is a wonderful and complete, no, a <u>perfect</u> statement of who our God is.

v Do trees pop into existence in your backyard? No. Do human babies suddenly appear in the nursery? No. Huge oak trees grow for single acorns about 5 grams in weight. Humans develop from single (bipartite) cells one would need a microscope to see. Astronomers have now concluded that galaxies originally developed from the gravitational singularities at the center of Black Holes by organizing the space particles within their gravitational reach. And, scientists have concluded that the entire universe was created with the explosion of the Cosmic Singularity, an initial singularity which was a point of infinite density and mass, thought to have contained the matter of the entire universe. What does all of this tell you about God the Creator, Jesus and how He went about Creation?

vi An example regarding the control of muscles in the body: There is DNA in every cell, except red blood cells. Each extremely thin strand of DNA is about 70 to 80 inches long. The DNA is wrapped 1.65 turns around bead-like proteins called histones. This assembly is further coiled into a form called a solenoid chromatin fiber, which has electrical characteristics. 23 pair of the solenoid chromatin fibers exist in the nucleus of the neuron.

When one decides to move an arm or any muscle in the body, the Particle responds by generating an electromagnetic Field Signal in the appropriate nuclei of motor neurons. The Field Signal induces a current in the solenoid chromatin fibers and action potentials (voltage spikes) are produced within the neurons which travel down the axons (electro-chemical conductors) to the appropriate muscles. See the following drawing.

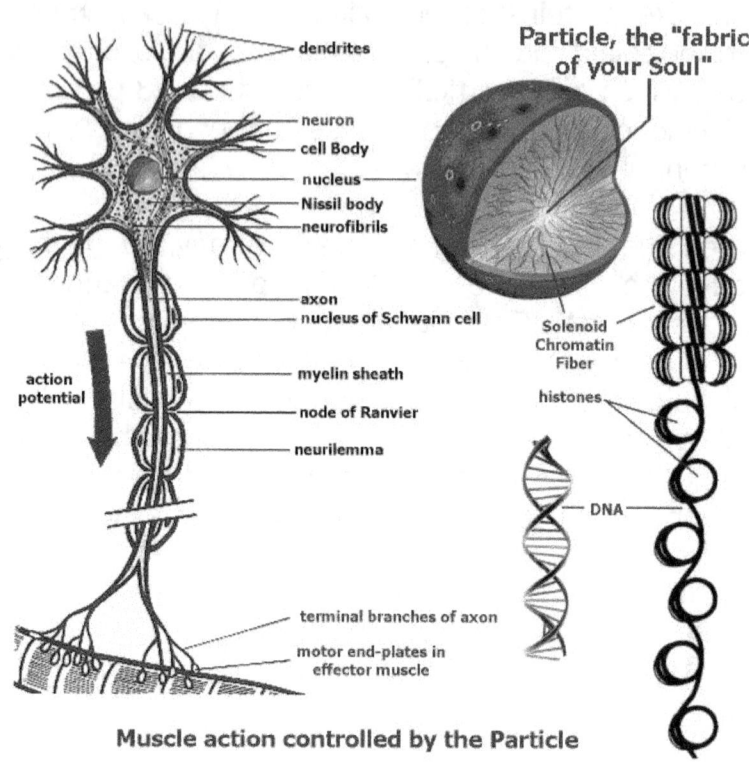

Muscle action controlled by the Particle

[vii] Some scientists, such as Dr. Rupert Sheldrake, who were looking for an explanation of everything, and especially how so much information could exist in a single cell at conception, have considered that the information necessary to develop the (body) of a new individual must come from a field that is surrounding the primary (stem) cell - a field that <u>somehow</u> has all of the information necessary and specific to the individual. They call it the morphogenetic field. They also believe it pervades all space, interacts with all matter and energy, and is even the basis of the Unifying Field which

Einstein was searching for. Some have suggested that the morphogenetic field interfaces with the electromagnetic field of the brain and is involved in how we can recall memories. Dr. Sheldrake goes on to say, "Morphogenetic fields are basically non-physical (spiritual?) blueprints that give birth to forms."

[viii] A 10-year-old boy is knocked unconscious by a baseball. Following that traumatic blow, he suddenly can do calendar calculations. He can also remember the weather, along with other autobiographical details of his daily life, from that time forward. An elderly woman who had never painted before becomes a prodigious artist after a particular type of dementia process begins and progresses. Another elderly patient with dementia has a similar sudden epiphany of ability, but this time in music. A 56-year-old builder, who had no particular prior interest or skills in art, abruptly, for the first time in his life, becomes a poet, a painter and a sculptor following a stroke that he miraculously survived. An 8-year-old boy begins calendar calculating after a left hemispherectomy for intractable seizures. These are examples of what I call the "acquired" savant, or what might also be called "accidental genius." Darold A. Treffert, MD, Wisconsin Medical Society.

[ix] "The prevailing wisdom of modern biology has it that cells are immensely complex, but rigidly operating chemical machines that derive their operating instructions internally from their genes and externally from chemicals and electrical signals emitted rigidly by other cells. Unable to believe that any machine can be designed that contains an instruction library which anticipates all the mishaps and glitches of a billion years

of evolution without crashing over and over again, I began almost three decades ago to search for signs that the cell was actually a 'smart' machine. In other words, I looked for experimental evidence that cells contained a signal integration system that allowed them to sense, weigh and process huge numbers of signals from outside and inside their bodies and to make decisions on their own." Guenter Albrecht-Buehler, Professor of Cell and Molecular Biology at Northwestern University. In other words, he does not believe that the human cell, containing DNA, can function without some form of operating intelligence. (The Particle.)

[x] If the Particle is at the functional center of every cell, as the Theory proposes, what happens when the organ of a living person, or that of a dying person, is transplanted into the body of another whose defective organ is removed? If recallable memories are acquired along with the organ by the recipient, as Dr. Pearsall has documented, then the transplanted organ retains the Particle of the donor. The Theory explains this phenomenon, but also opens the question, "should transplants take place?" The memories "transplanted" are limited and temporary, or Dr. Pearsall would have said otherwise.

[xi] A 47 year-old white male foundry worker, who received the heart of a 17 year-old black male student, discovered after the operation that he had developed a fascination for classical music. He reasoned that since his donor would have preferred 'rap' music, his newfound love for classical music could not possibly have anything to do with his new heart. As it turned out, the donor actually loved classical music, and died

"hugging his violin case" on the way to his violin class.

xii Do you remember playing on a freshly cut lawn? Did you ever run your hand over the top of the grass? Remember the feeling you got on the palm of your hand? Extend out your arm right now with the palm of your hand facing down. Move it back and forth and think about how it felt when you touched the grass. Can you feel the blades of grass touching your palm? How does that happen?

xiii The basic concept presented by Dr Michael Behe is that of 'irreducible complexity'. A system that is irreducibly complex is one in which precise components work jointly to perform the basic function of the system. It also means that if any part of that system were to be absent or removed, the system would cease to function. Therefore, any step to simplify an irreducibly complex system would result in a non-functional system. This presents an insurmountable problem for the Darwinist. If one alleges that all systems evolved by gradual addition to previously functioning systems, how does one explain a complex system that would not perform its basic function if it were missing even a single component?

xiv The refresh rate of a television represents the number of times per second that the image is flashed or refreshed on the screen. By flashing a series of still images, the TV (just like film or animation) creates the illusion of motion. The refresh rate is measured in hertz (named after physicist Heinrich Hertz, who did groundbreaking work on electromagnetism). So if a Refresh rate is 120 Hz, it means that the image is refreshed 120 times per second. In theory, the more pictures per second, the more realistic the motion or video should appear. [Jason Padjett, from Tacoma, Washington, received a severe blow to the back of the head and afterwards, he describes his vision as "discrete picture frames with a line connecting them, but still at real speed. If you think of vision as the brain taking pictures all the time and smoothing them into a video, it's as though (one) sees the frames without the smoothing."]

[xv] Nerve impulses are extremely slow compared to the speed of electricity, where electric current can flow on the order of 50–99% of the speed of light; however, it is very fast compared to the speed of blood flow, with some myelinated neurons conducting at speeds up to 120 m/s (432 km/h or 275 mph). Nerve conduction velocity – Wikipedia

[xvi] I read in an article on the process of healing that fibroblasts are cells key to the process. Quoted from the accompanying article: "The principle function of fibroblasts is to maintain the structural integrity of connective tissues. Fibroblasts have diverse appearances depending on their location and activity." Interesting. The article continues, "Fibroblasts can often (somehow) retain positional memory of the location and tissue where they had previously resided. It is fibroblasts and related connective tissues which sculpt the 'bulk' of an (injured) organism." Fibrocytes are newly identified cells that can rapidly and specifically migrate into the site of tissue injury at the time of injury. They are derived from bone marrow. Just what directs and controls the actions of fibroblasts and fibrocytes?

DNA is not even mentioned in this article. Could it be the soul as suggested by the Theory? I think so.

Quotations above from C. Michael Gibson M.S., M.D., Associate Professor of Medicine, Harvard Medical School.

[xvii] "Moist wound healing is the name given to the observation that a wound that is kept optimally moist will have better outcomes than one that is allowed to dry out. Studies have shown that a moist wound heals between three and five times faster than a dry wound.

The idea of moist wound healing was first defined during the 1960s. During this time, early pre-clinical and clinical research conducted by the British pioneer, George D Winter, first demonstrated the benefit of a moist environment in optimizing wound healing.

Our understanding of moist wound healing has come a long way since the days of the first research of Dr Winter. We now understand why wound healing is promoted by a moist environment. This is due to several, parallel processes. Firstly, by preventing scab or crust formation over the wound bed, a moist wound environment eliminates the energy and time that would have been required for the body to break down these materials. Keratinocyte-travel time and distance across the wound surface are also greatly reduced, as the cells are able to easily migrate across the moist wound bed rather than burrow underneath the wound bed to find a moist area upon which to move forward. A moist environment also traps enzymes within the wound bed, facilitating autolytic debridement. And finally, a moist wound environment preserves growth factors within the wound fluid and increases fibroblast proliferation and collagen synthesis."

Posted on Wound Education Social Network by Laurie Swezey, RN, BSN, CWS, CWOCN.

[xviii] Institute of Cancer Research: "We provided the first conclusive evidence that the basic cause of cancer is damage to DNA. The discovery changed scientific opinion dramatically and marked a turning point for cancer research. Until that point, scientists had assumed carcinogens caused cancer by acting on proteins, rather than genes.

Research conducted here at The Institute of Cancer Research in the 1950 and 60s provided the first conclusive evidence indicating that damage to DNA is the root cause of cancer.

Cancer cells replicate at an accelerated rate, often ignoring the normal controls on cell division and growth. Proteins within cells regulate growth and division, and were widely assumed to be the targets for cancer-causing chemicals. However, research conducted at the ICR by Professor Philip Lawley and Professor Peter Brookes showed that cancer is caused by damage not to proteins but to DNA."

[xix] My brother called to tell me he was diagnosed with myelofibrosis. Some, not all, of the stem cells of his bone marrow have mutated to divide into cells that produce excessive proteins consisting of tiny fibers. That's the "fibrosis" part. Lower red blood cell

production in his bone marrow resulted. Some of the blood cell production was picked up by his spleen, which caused it to swell. The problem, in my non-medical opinion, is that errant stem cells in his bone marrow are having difficulty making division decisions regarding the type of cells to become – poor communications with the Particle. I told him to ask his doctor if he can make the mutant cells healthy. The doctor gave him hydroxyurea instead, a drug which kills fast dividing cells (and some normal cells). My brother died from the illness.

[xx] "Do the eyes scan dream images during rapid eye movement sleep? Evidence from the rapid eye movement sleep behavior disorder model." A study by Laurène Leclair-Visonneau, Delphine Oudiette, Bertrand Gaymard, Smaranda Leu-Semenescu and Isabelle Arnulf, published in *Brain, a Journal of Neurology* – May 16, 2010.

The Catholic Catechism compared to the Atemporal Particle Theory

355 "God created man in his own image, in the image of God he created him, male and female he created them." Man occupies a unique place in creation: (I) he is "in the image of God"; (II) in his own nature he unites the spiritual and material worlds; (III) he is created "male and female"; (IV) God established him in his friendship.

More accurately stated:

355 Jesus created mankind in his own image, in the image of God he created them, male and female he created them." Mankind occupies a unique place in creation: 1) they are "in the image of Jesus who is God"; 2) in their own nature they unite the spiritual and material worlds; 3) they are created "male and female"; 4) Jesus established mankind in his friendship.

All of the following have been "corrected" within parenthesis.

360 . . . in the unity of its nature, composed equally in all men (and women) of a material body (a spiritual body) and a spiritual soul; in the unity of its immediate end and its mission in the world; in the unity of its dwelling, the earth, whose benefits all (mankind), by right of nature, may use to sustain and develop life; in the unity of its supernatural end: God himself, to whom all ought to tend; in the unity of the means for attaining this end . . . in the unity of the redemption wrought by Christ for all.

362 The human person, created in the image of (Jesus), is a being at once corporeal and spiritual. The biblical account expresses this reality in symbolic language when it affirms that "then the LORD* God formed (a) man of dust from the ground and breathed into his nostrils the breath of life;* and the man became a living being." (Mankind), whole and entire, (was) therefore *willed* by God.

* Note: Jesus blew into Adam's nostrils to fill them with oxygen for the first time as one would a newborn baby who was not breathing.

363 In Sacred Scripture the term "soul" often refers to human *life* or the entire human *person*. <u>But "soul" also refers to the innermost aspect of man, that which is of greatest value in him, that by which he is most especially in God's image: "soul" signifies the *spiritual principle* in man</u>.

The underlined portion of 363 is totally wrong, not fixable.

364 The human body shares in the dignity of "the image of (Jesus who is God)": it is a human body precisely because it is <u>animated by a spiritual soul</u>, and it is the whole human person that is intended to become, in the body of Christ, a temple of the (Holy) Spirit:

Note: "animated by a spiritual soul," just as the Atemporal Particle Theory teaches. This is also in 466.

* Jesus is LORD.

365 The unity of soul and body is so profound that one

has to consider the soul to be the "form" of the body: i.e., it is because of its spiritual soul that the body made of matter becomes a living, human body; spirit and matter, in man, are not two natures united, but rather their union forms a single nature.

The Catholic Church has confused the soul with the spirit. St. Paul in 1 Thessalonians 5:23 got it right.

More accurately stated:

365 The unity of soul and body is so profound that one has to consider the soul to (contain) the "form" of the body: i.e., it is because of its spiritual soul that the body (both corporeal and incorporeal) become a living, human body; spirit and matter, in man, are not two natures united, but rather their union forms a single nature.

366 The Church teaches that every spiritual soul is created immediately by God - it is not "produced" by the parents - and also that it is immortal: it does not perish when it separates from the body at death, and it will be reunited with the body at the final Resurrection.

Statement 366 is too wrong to correct, although it would be correct to say, the spirit and soul are both immortal and do "not perish when (they) separate from the body at death and (they) will be reunited with (a new) body at the final resurrection (i.e., at the creation of the new heaven and new earth)."

367 Sometimes the soul is distinguished from the spirit (in the Bible): St. Paul for instance prays that God may

sanctify his people "wholly", with "spirit and soul and body" kept sound and blameless at the Lord's coming. <u>The Church teaches that this distinction does not introduce a duality into the soul. "Spirit" signifies that from creation man is ordered to a supernatural end and that his soul can gratuitously be raised beyond all it deserves to communion with God.</u>

The apostle Paul was correct. The Church (underlined) is obviously unable to explain Paul's statement.

461 Taking up St. John's expression, "The Word became flesh," the Church calls "Incarnation" the fact that the Son of God assumed a human nature in order to accomplish our salvation in it.

464 The unique and altogether singular event of the Incarnation of the Son of God does not mean that Jesus Christ is part God and part man, nor does it imply that he is the result of a confused mixture of the divine and the human. He became truly man while remaining truly God. Jesus Christ is true God and true man.

466 . . . (Jesus) uniting to himself in his person the flesh <u>animated by a rational soul</u>, became man." Christ's humanity has no other subject than the divine person of the Son of God, who assumed it and made it his own, from his conception. For this reason . . . Mary truly became the Mother of God by the human conception of the Son of God in her womb: "Mother of God, not that the nature of (Jesus) or his divinity received the beginning of its existence from the holy Virgin, but that, since the holy body, animated by a rational soul, which (Jesus) united to himself . . . was born from her, Jesus is said to be born according to the flesh.

The Catechism has a lot of trouble explaining the above because the writers are unaware of Jesus being in the Concurrent State, unlimited by time and the sequence of events. Before time began, Jesus decided Mary would be his mother.

470 . . . For by His incarnation (Jesus) united Himself in some fashion with every (human). He worked with human hands, He thought with a human mind, acted by human choice and loved with a human heart. Born of the Virgin Mary, He has truly been made one of us, like us in all things except sin.

472 This human soul that the Son of God assumed is endowed with a true human knowledge. As such, this knowledge could not in itself be unlimited: it was exercised in the historical conditions of his existence in space and time. This is why the Son of God could, when he became man, "increase in wisdom and in stature, and in favor with God and man", and would even have to inquire for himself about what one in the human condition can learn only from experience. This corresponded to the reality of his voluntary emptying of himself, taking "the form of a slave".

481 (Jesus) . . . possesses two natures, one divine and the other human, not confused, but united in the one person . . .

Just like humans, each with a Sequential State body and a Temporal State body both with one soul but united as one person.

482 Christ, being true God and true man, has a human

intellect and will, perfectly attuned and subject to his divine intellect and divine will, which he has in common with the Father and the Holy Spirit.

Not true. Though fully God and fully man, Jesus was not "perfectly attuned and subject to his divine intellect and divine will, which he has in common with the Father and the Holy Spirit." He was focused entirely on his physical nature as we all are but relied on his intuition to know what the Father told him to say. This was true until his bodily death on the cross.

Theologians, like philosophers, try to explain everything with logic, having no foundational science behind their explanations.

Cancer and the Atemporal Particle Theory

We know that the Particle is at the functional center of every cell including cancer cells.

We know the Particle controls the functions of all cells through the DNA (within Solenoid chromatin fibers) in the nuclei of the cells by means of electromagnet radiation ("Field Signals") produced by the Particle.

We know that DNA within solenoid chromatin fibers acts as an high frequency, broadband (fractal) antenna (Department of Physiology, Columbia University agrees).

We know that the DNA of a cancer cell is damaged in some way and therefore does not receive, recognize or carryout proper instructions from the Particle regarding shape, purpose, color, structure, location or division limits. The difference between a benign tumor and a malignant tumor is that although both are not receiving division limits, the malignant tumor is also not receiving location instructions.

We know that "Low-frequency rTMS (repetitive transcranial magnetic stimulation) interferes in with the Field Signals from the Particle sent to "a localized area of the cerebral cortex, thereby (effectively) creating 'virtual lesions' (resulting in savant abilities in test subjects)." (Quote from PascualLeone et al 1999; Walsh and Cowey 2000; Hoffman and Cavus 2002).

We know that radiation therapy works by making many small breaks in a cell's DNA. These breaks keep cancer cells from growing and dividing and cause them to die. (American Cancer Society agrees). In other words, radiation therapy makes it impossible for the DNA to receive information at all from the Particle and the affected cells atrophy and die.

We know that x-ray radiation can partially damage DNA in cells which may result in the creation of cancer cells.

We can assume, therefore, that if we interfere directly with the Field Signals, produced by the Particle within the cells of a cancerous tumor, the cells will no longer receive any instructions from the Particle and will atrophy and die.

We know that, radiation therapy uses high-energy particles or waves, such as x-rays, gamma rays, electron beams, or protons, to destroy or damage the DNA in cancer cells. This radiation causes unwanted damage to the DNA of other cells in front of or near the cancer cells.

We know that Field Signals from the Particle are high frequency but also of extremely low energy.

If instead of radiating cancer cells with high-energy from outside of the body to physically damage DNA, a way should be found to interfere with the Field Signals from the Particle within the cells of a tumor, thereby causing the cancer cells to atrophy and die without destroying adjacent cells.

A low energy transmitting device, tuned to the same frequency as the Field Signals from the Particle, incorporating disruptive modulation, and focused on the area of the cancerous tumor, should be able to override the Field Signals from the Particle.

"There is emerging evidence that the growth of cancer cells may be altered by very low levels of electromagnetic fields (27.12 MHz) modulated at specific frequencies." (quote from the Hospital das Clínicas da Faculdade de Medicina, University of São Paulo)

"There is clinical evidence that very low and safe levels of amplitude-modulated electromagnetic fields . . . may elicit therapeutic responses in patients with cancer. However, there is <u>no known mechanism explaining the anti-proliferative effect</u> of very low intensity electromagnetic fields." (quote from the British Journal of Cancer)

"The mechanism by which AM RF EMF have direct antiproliferative effect and disruption of the mitotic spindle on cancer cells <u>is largely unknown</u>." (quote from Zimmerman JW, et al. *Br J Cancer.* 2012;106:307-313.)

A medical startup ("Startup") in Germany, has developed a low energy transmitting device, tuned to 27.12 MHz with amplitude-modulation at a specific frequency for cancer therapy.

Startup believes that "a specific calcium channel, Cav3.2, (is) acting like an antenna for the radio signals they send out, which allows calcium to penetrate the cancerous cell membrane and go into the cell, triggering the growth arrest of the cell." (quotation is from the Startup)

DNA is known to be a broadband antenna, designed to receive instructions from the cell's Particle. When the DNA of the cell is damaged, it cannot not receive, recognize or carryout proper instructions from the Particle regarding shape, purpose, color, structure, location or division limits and it thereby may become a cancer cell.

Matching the amplitude modulation frequency of the RF to the damaged DNA of the cancer cell is not mentioned in Startup's reports, but that is what is happening. It is the damaged DNA that is resonating with the amplitude modulation frequency resulting in the overriding of the Field Signals from the Particle "(allowing) calcium to penetrate the cancerous cell membrane . . . triggering the growth arrest of the cell."

Startup has found a way, through empirical tests, to match the modulation frequency of the 27.12 MHz transmission with the damaged DNA, or that part of the damaged DNA, which if not so damaged would not otherwise be able to resonate with the modulation frequency.

The cancer-identifying amplitude modulation frequency of the RF is said by Startup to be specific to the particular cancer type to be treated, but Startup's own report does not fully bear this out. However, since cell structures and DNA content vary from cell type to cell type, it is possible that certain damage, due to radiation for example, to the DNA of the various cell types may be repeatable, thus causing Startup's conclusion that they can target certain cancers.

It would be important for Startup to know that they are not targeting cancer cell types (liver, breast, pancreas, etc.), but that they are targeting damaged DNA within cancer cells, which may be similarly damaged in specific cell types, and what their device is really doing is interfering with information coming from the Particle which otherwise might in fact be directing the cancer cells to divide and even permitting their relocation elsewhere within the body.

© January 2020 – John Beiswenger

Afterword

I have placed all of the subject-related information I currently have into this book. Although it may be quite difficult and too much to read, what I want you to take away from all of this is not that impossible to grasp.

> There is a single, massless, atemporal Particle at the functional center of every cell in the body. It has been present since conception, being the combined Particles of the parents. It is the source of cellular control; the repository of memory and thereby facilitator of consciousness.

In my thirty-years of research, I have found no other plausible answer, and none has been suggested to me by other scientists and authors. This is why I had to write this book.

> The discovery of the existence and purpose of the Particle will surpass in importance the discovery of the structure of DNA.

I would welcome, and work with, any scientist willing and able to pursue the search. It will lead to a more fully understanding of vision, consciousness, autism, savant abilities,* cellular control and even cancer, its causes and treatment.

You can contact me through my author website at JohnBeiswenger.net.

© 2020 John Beiswenger
Christian, Author, Engineer

* "Until we can explain the savant, we cannot fully understand the brain or human potential."
The Kavli institute for Brain and Mind.

Acknowledgements

My primary acknowledgement must go to my beautiful wife, Kim, who not only encouraged me in my research, but actually asked me to write this book. Kim has frequently found for me important articles relating to The Atemporal Particle Theory.

My second acknowledgement goes to my older son, David, a degreed business-software marketing executive. He has often sent me articles relating to the Theory, and he purchased and sent me Jeff Hawkins book, On Intelligence, which I have cited frequently.

Joe Johnson, a biochemist and Jim Wilson, an MD, spent hours listening to me discussing the development of the Theory. We worked together over the past 12 years identifying the pre-symptomatic signs of infection detectable by the medical device on Page 5.

www.ingramcontent.com/pod-product-compliance
Lightning Source LLC
Chambersburg PA
CBHW070634220526
45466CB00001B/168